FLUID NEW YORK

FLUID NEW YORK

cosmopolitan urbanism and the green imagination

DUKE UNIVERSITY PRESS : *Durham and London* : 2013

Printed in the United States
of America on acid-free paper ∞
Designed by Amy Ruth Buchanan
Typeset in Quadraat by Tseng
Information Systems, Inc.

Library of Congress Cataloging-in-Publication Data
Joseph, May.
Fluid New York : cosmopolitan urbanism and the
green imagination / May Joseph.
pages cm
Includes bibliographical references and index.
ISBN 978-0-8223-5460-4 (cloth : alk. paper)
ISBN 978-0-8223-5472-7 (pbk. : alk. paper)
1. New York (N.Y.)—Civilization—20th century.
2. New York (N.Y.)—Civilization—21st century.
3. Cosmopolitanism—New York (State)—New York.
4. Urban ecology (Sociology)—New York (State)—
New York. I. Title.
F128.55.J67 2013
974.7′1044—dc23 2013010099

contents

illustrations

Figures

Maps

acknowledgments

I am indebted to many strangers and friends who made this book possible. My wonderful, patient editor, Ken Wissoker, remained a constant support through the twisted journeys of the writing of this book. To Smriti Srinivas, Brian McGrath, and Victoria Marshall I owe much for long conversations on ecology, water, and cities, conversations that were the bedrock of this excursion. Immeasurable thanks are due George Lipsitz, Sandra Richards, Mark Nowak, Smriti Srinivas, Brian McGrath, Randy Martin, and Geetha Varadarajan for their rigorous comments on earlier drafts of the manuscript. Kanishka Goonewardena, Abidin Kusno, Beng-Lan Goh, Rodrigo Tisi, and Dorita Hannah offered a fruitful context for my queries into urbanism. My colleagues Jon Beller, Ira Livingston, Sameetah Agha, Greg Horowitz, Uzma Rizvi, Carl Zimring, Francis Bradley, Josiah Brownell, Amy Guggenheim, and Thad Ziolkowski provided a generous, supportive environment. Thank you all. Conversations with many scholars, including Rick Bell, Grahame Shane, Susan Henry, Una Chaudhuri, the Conjunctures Collective, Markus Vink, Pedro Machado, Patricia Clough, Percy Hintzen, Jean Rahier, David Lelyveld, and Pia Lindman, fueled my writing. Many thanks for the creative commiserations of James Cascaito, Meena Alexander, David Van Leer, Lisbeth During, Nandini Sikand, Gloria Zimmerman, Chris and Diane Power. To Alana Free and Marjorie Tesser I owe much for detours in writing. Carol Gaskin's skillful editing and Jeffrey Canaday's experienced hand gave final shape to the book. Leigh Barnwell,

Amy Buchanan, and Susan Albury, thank you for shepherding the manuscript into print. I am especially indebted to my cartographer, Daniel Hetteix, for his meticulous fact-checking and indispensable cartographic renderings that bring this book to life. To Sanjay Subrahmanyam, a special thanks for introducing me to perspectives on oceanic histories. I am also greatly indebted to A. J. James and Nitya Jacob for patiently enlightening me about global water politics. Thanks are due to Dean Toni Oliviero, Dean Andrew Barnes, and the Pratt Mellon Grants, as well as the Faculty Development Fund Grants of 2004 and 2009, for funding this research. Special thanks are also due to Thomas Bender, who allowed me to sit in on the seminars on cities that he organized at the Center for Advanced Studies at New York University. The extraordinary environment provided the foundation for this book.

To my anonymous reviewers, a big thank you for your thoughtful and detailed edits, which transformed the manuscript. Special thanks are also due Nancy Kandoian, of the New York Public Library's Map Room, who so patiently assisted me in my research through the years.

I owe much to my parents, intrepid urbanists, for their indefatigable curiosity in walking cities. Special mention must be made of two little girls, Tori and Amelia Power, for roaming this fair city with my daughter through snow, sludge, rain, and shine, researching carnivals, protests, performances and the waterfront, year in and year out. Finally, to my husband Geoff and daughter Celine, my deepest gratitude for patiently living under the shadow of this book, and surviving Hurricane Sandy with humor.

preface

September 11, 2001

Billowing smoke. Scorching flames interrupt the Manhattan skyline on a stunning morning in the waning days of summer, 2001. In an unspeakable urban catastrophe, the World Trade Center goes up in flames. A surreal theatricality holds the city spellbound. People pray, weep, howl, scream, tremble, gasp. Symbols of late twentieth-century economic modernity, the raging goliaths stand in soaring fury against a smoldering backdrop—until, unimaginably, the impossible beauty implodes and the 1970s behemoths collapse like giant matchboxes, a rubble of burnt lives.

I am transfixed as the horror swiftly undulates around my one-month-old, the shudder of the impact rattling her body and shattering the city's. How could the Twin Towers possibly disappear before our eyes? An impressive madness, precise, strategic, and inconceivable, the soul-wrenching cataclysm tugs at the core of what it means to live in a city. The quivering ground and ricocheting thud send tremors of doubt through the urban landscape, forcing a new consideration of how people cope with the vulnerability of living in a dense metropolitan center.

The uncertain and informal improvisations of an island city gripped in an unfolding chaos that was both gigantic and intimate arose amid the pandemonium of curbs, corners, streets, parks, subway tunnels, bridges, highways, waterways, and public spaces. In that space of shock, a tremendous performance of civic commitment was staged in myriad provisional ways, the most remarkable assertion that cities are powerful lived events, composed of the symbiosis of human biopower and the built hardscape.

A stranger's voice gently assures us that this too will pass.

prologue

As *Fluid New York* was rolling into press, a "superstorm" of unimagined proportions surged the eastern seaboard, deluging large portions of New Jersey and New York's coastline. Hurricane Sandy's ferocity caught New York City unprepared for its scale and devastation. A catastrophic scenario in which entire neighborhoods were destroyed by flooding and uncontrollable fires swept the city. Random explosions, including that of the primary source of power to downtown Manhattan, the Con Edison station, and raging fires in Breezy Point, Queens, generated considerable anxiety. New York's major airports, La Guardia and JFK, were waterlogged, halting all flights to New York for nearly a week. Entire waterfront neighborhoods, Brighton Beach, Far Rockaway, Breezy Point, Manhattan's downtown, and large swaths of Staten Island, were flooded.[1] New York City's low-lying areas and coastal New Jersey were unprecedentedly submerged by seawater. This chaotic scene was exacerbated by a blackout lasting five days in downtown Manhattan and parts of Brooklyn, Queens, and Staten Island. In some areas, an entire month passed before the restoration of power and water.

Hurricane Sandy decimated New York City. The storm plummeted the city into a state of infrastructural breakdown and shock. Subway networks incurred extensive water damage. Sewage plants were flooded, causing seepage and large-scale contamination in many neighborhoods, as well as the city's rivers. Many outlying neighborhoods went without food, electricity, water, or amenities for weeks. Entire parking lots of cars, storage depots for art galleries and fashion businesses, and research laboratories for major hospitals were flooded, sometimes with up to nine feet of water in basements.[2] Along the West Side Highway in Manhattan and other low-lying areas, high-rises suffered extensive flooding and water damage, making it difficult for residents to occupy their homes for weeks. Damage to major transportation hubs like the Brooklyn-Battery Tunnel paralyzed the city. The fates of many small businesses and individual homes were sealed by mold and saltwater damage, causing severe personal hardship. Hospitals had to evacuate the critically ill and close their entire operations in downtown Manhattan, greatly reducing the availability of hospital beds for New York's injured, and the mentally ill. As of the end of January, hospitals in downtown Manhattan had yet to resume their full functioning. Entire communities from Queens, Brooklyn, and Staten Island were physically displaced and reduced to migrant existences because of extensive water damage to their homes. These cursory images capture only a glimpse of the devastation waged on New York City by the storm.

Hurricane Sandy marks a major turning point in New York City's history, equivalent in seismic shift to the impact that 9/11 wrought upon the city. It materially affected a much greater population over a larger area in the Northeast, causing damage amounting to tens of billions of dollars. The massive storm surge has permanently changed New York's relationship to its waterfront, and the ensuing debates on climate change. This point was underscored by Mayor Bloomberg's comment: "We are not going to abandon the waterfront. . . . But we can't just rebuild what was there and hope for the best. We have to build smarter and stronger and more sustainably."[3] Bloomberg captures some of the biggest challenges facing New York City's recovery from what meteorologists have noted was a storm of anomalous proportions. Bloomberg's public endorsement of Obama in the presidential race, to mark the importance of global climate change for coastal cities like New York, comes on the heels of the first cautious, public acknowledgment by government officials in the United States that the rise of global warming is dramatically altering our urban landscape.

Hurricane Sandy was a storm of unprecedented force in New York's documented history. Its influence on the city's urban imagining is only beginning to be fathomed. One thing is certain: there is a before and an after Hurricane Sandy. Before Sandy, New York's position on disaster prevention was largely at the level of designated commissions in Albany and policy discussions, without the political will to implement large-scale planning that such an endeavor would entail. After Sandy, formerly guarded government positions have been replaced by vociferous articulations by government entities at various levels, to address difficult questions regarding the effects of the inevitable, grim future of rising oceans and changing weather patterns on urban life. After Mayor Michael Bloomberg, Governor Andrew Cuomo, of New York, Governor Christie, of New Jersey, President Clinton, and President Obama have all publicly noted the challenges of disaster recovery.[4] Superstorm Sandy has forced urban Americans to talk about sustainable climate-change management and its impact on cities.

At a local level, New Yorkers now realize they live in an island city, whose extensive shoreline exists in a delicate ecological balance between man-made waterfront and rising ocean levels. For the first time in the city's recent history, New York is culturally awake to the reality that its archipelagic structure creates a way of life that is precariously dependent on changing weather patterns. The waterfront, once a site of maritime industry, and most recently a destination of leisure, has emerged a vulnerable coastline whose futurity is once again dramatically being renegotiated, this time by saltwater damage, mold, fear, loss, misery, tragedy, and pernicious real estate developers.

Hurricane Sandy has exposed New York as a city shockingly under-prepared for its ecological future as a coastal city. Despite its First World status as a city of style, of cosmopolitan urbanity, New York deplorably lags behind many European and non-Western cities like Rotterdam, in Holland, and Singapore, when it comes to storm surge planning.[5] New York had not fully grasped the reality of its coastal location till the catastrophe of Sandy. Failing to invest in disaster prevention after Hurricane Irene, in 2011, New York was unprepared for the extent of the storm surge and incurred catastrophic water damage amounting to billions of dollars and considerable human loss. Cities by the sea have to prepare for a future of rising ocean levels. For New York, it was too little too late. What lessons can New York learn from other coastal cities?

The Future of Vertical Living

New York is always learning from itself. However, as Hurricane Sandy has proven, it is not learning fast enough. This book emerged as an investigation into the future of vertical living when the waterfront emerged once again as the city of the future. What forms of urban arrangements, social intimacies, and infrastructural developments was America's densest city contending with during this new, unprecedented spurt in waterfront revitalization? This was a question that interested me.

The scenario that emerged by the end of the 1990s and lasted until 2012 involved a steady embrace of New York's waterfront as the new gold coast of the city. Tall, gleaming residential towers rose along Brooklyn's once low-level, industrial waterfront, competing with Manhattan's older, more contained distribution of high-rises. Reclaimed land on either side of the East River was built higher and bolder than ever before. The future of vertical living in New York seemed to be in a renaissance of furious upward development. The maritime sky of Manhattan was once again being shaped by bold skylines along its shores. What kinds of philosophies of the city, social imaginings, and cultural negotiations was New York grappling with during this era of vertiginous growth along its shoreline? Whatever the answer, Hurricane Sandy has permanently altered those preoccupations.

Sandy has washed ashore new implications for vertical living in coastal cities. High-rise living in the grip of a total blackout, without generators and the accompanying safety measures to prevent total electrical outage, has presented a disturbing scenario for the future of storm-bound cities like New York. Living in density today has come to mean being prepared for climate-related emergencies, including heat waves, electric blackouts, the deprivation of water, heat, and social services, and the sudden threat of flooding. This shifting scenario of dense living has introduced a new historical understanding of what it means to live in a vertical city—it means one must expect extreme scenarios and prepare for them as part of the commitment to live in coastal metropolitan centers like New York. The realization that one has to be prepared for catastrophe is slowly emerging from between the bookend experiences of 9/11 and Hurricane Sandy.

Catastrophe in New York is its own peculiar experience. It is shaped by the height, scale, and proximity of human dwellings. The stories emerging from Queens, Brooklyn, and Staten Island about the impact of the blackout on people's lives are myriad and diverse. In Staten Island, illegal immigrants without flood insurance whose homes were devastated by flooding live in inhospitable conditions, without running water and toilet facilities. In the Far Rockaways, elderly residents spoke of being stranded on a thirtieth floor without food, water, electricity, and in some cases, medications.[6] In Brooklyn, residents described the flooding of sewage water into their homes because pipes from the nearby sewage plant exploded.[7] People living in downtown Manhattan spoke of a renewed sense of community and the tangible emergence of an ecological vulnerability. Perhaps the most striking incongruence came from the harrowing stories of the daily lives of those servicing Manhattan's casually affluent uptown, as residents in the Upper West Side settled into a comfortable week of unexpected vacation because of the storm. The ease of Manhattan's Upper West Side in the wake of the utter chaos of the rest of New York heightened the dramatic physical difference between the darkness below Thirty-Ninth Street, in Manhattan, and the brightly lit avenues above.[8] It was the acute sense of two topographies, two cities. The brightly lit uptown also accentuated an extreme sense of the ecology of the island, that of the low-lying, vulnerable downtown, flooded and paralyzed by the blackout, and a protected uptown, on higher ground and for the most part protected from the worst ravages of the storm, and therefore considerably unscarred by the trauma of surviving Hurricane Sandy.

South of Fourteenth Street, in Manhattan, a carnival-like atmosphere and impromptu social gatherings outside cafés offered temporary respite, as the blackout extended indefinitely. Neighbors assisted the elderly in darkened stairwells, and transient friendships were struck as people helped each other through the difficult dark days without electricity or water. Stalwart restaurants like the West Village mainstay Bonbonnier kept their doors miraculously open twenty-four hours a day, serving coffee and hot food, without electricity or a water supply, to desperately cold and hungry residents. In Queens and Brooklyn, residents reported the kindness of neighbors and strangers stopping to check in on the elderly and the sick stranded in high-rises in waterlogged regions of Far Rockaway and Brighton Beach. The tragic scenario of the once cozy and enviable community of Breezy Point being wiped out by the ferocity of the storm remains the poster child for Hurricane Sandy, the neighborhood's picturesque beach homes burnt to the ground by tongues of fire.

For the past decade, New York has been feebly talking about disaster preparedness without putting real money behind an effort to build sea walls and storm surge protection. Instead, the city has leaned toward disaster relief, creating extensive loss of life and hardship as a result. The future of vertical living in New York is in the balance. How we New Yorkers plan for the next storm, and how we prepare our residents for the challenges of living in vertical cities under threat of power outages caused by sudden flooding or extreme heat are some of the planning issues that will shape the future of metropolitan life in coastal areas. As after 9/11, New

York is once again in salvage mode, this time recovering from the devastation of a natural disaster. How it learns from its unpreparedness, and how it harnesses its technologies and expertise to structurally anticipate the next weather disturbance related to climate change, will determine the city's future.

introduction

New York is a place of passions. It demands extreme reactions. When I first moved to New York City from Santa Barbara, California, in 1992, to take up my first job, a friend from Brooklyn said to me cryptically, "New York is a commitment." I grew up in Dar es Salaam, a laid-back coastal city on the Indian Ocean, and had lived in other cities before coming to New York. I didn't quite understand why New York would be more overwhelming than other world cities.

Life in New York is indeed different from a life in other cities. The city is challenging, even ruthless. It is also intimate and compassionate. Observing the peculiar culture of New York and the city's barometers of belonging over the years, I have come to understand the commitment it takes to live in New York City, as opposed to passing through it. It is the difference between cultural cosmopolitanism and political cosmopolitanism. To pass through New York is to participate in its multiplicity without having to deal with the limits, restrictions, and disciplinary boundaries through which cultural difference must find mutual tolerance and common cause. To immerse oneself in the daily life of New York is to smell stale garbage in August and grapple with the competing inequities of cosmopolitan urban citizenship. I began to write about this indomitable New York of fraught, transnational human flows.

What is it, I found myself asking, that makes citizens of New York simultaneously more provincial and more cosmopolitan? And does the corporeal, experiential practice of living in New York today have any new lessons to offer the furiously expanding world of global urbanisms? New Yorkers are vociferous, active urban dwellers with strong opinions about everything. They comprise people of every hue and stripe relating to each other on a daily basis. As a newcomer, I found that the peculiar public culture of New York, a culture of cosmopolitan becoming, had shifted my perceptions of city life.

To cultivate the urban dweller into a cosmopolitan subject, an elaborate set of historical, political, institutional, and legal frameworks have developed in New York over four hundred years to actively constitute that locally bound yet cosmopolitan citizen known as the New Yorker. One particularly noteworthy characteristic of New York cosmopolitanism is the responsible participation of its denizens in the shaping of their city's identity as both a small town and a world city. Such social techniques of urban participation have a pedagogical effect, imperceptibly transforming other residents of the city into cosmopolitan citizens. This book takes a closer look at some of these quintessentially urban, expressive techniques of city-making, to better understand the elusive processes of cosmopolitan citizenship.

New York is not just about heterogeneity, a quality other metropolises share. Neither is New York's cultural claim to cosmopolitanism particularly unique at this time in history. Other world cities exhibit similar signs of cultural cosmopolitanism, with fragments of different world cultures constituting the visual fabrics of their urban façades. What is worth studying about New York is its rigorous, democratic machine of political cosmopolitan citizenship. Becoming a New Yorker is not just about the consumption of multicultural sensations in a globally mediated environment. It is more challengingly about a set of performative engagements that activate and promote mutual respect and coexistence within the finite spaces of the metropolis. This socially contestatory, culturally volatile, yet rational endeavor is ultimately a utopian undertaking in political citizenship. It makes New York City endlessly interesting to write about.

This book arose amid the fast-paced changes and hard-to-swallow realities that shaped millennial New York City. Each chapter describes an immigrant's journey into a slice of metropolitan life centered around historic downtown Manhattan. My driving preoccupation is the future of dense living. Wide-eyed and low to the ground, writing from the vantage point of a city dweller cycling to work from Manhattan to Brooklyn, or swept suddenly into an urban spectacle, I take a frog's perspective on a city at a difficult crossroads in its history.

My entry into New York City coincided with the city's dramatic reinvention of its self-imagining as an island city with a working waterfront fallen into disrepair. A public culture of green thinking with an investment in sustainability and water ecology emerged in the city's consciousness during this period. The practice of rethinking New York's past as a maritime city also emerged on a scale unprecedented since Jane Jacobs's and Robert Moses's era. Learning from cities like Curitiba, Copenhagen, and London about how to green its environments, New York is in the process of reconstituting itself as an ecological city.

A singular question is pursued in this book: What is the vision shaping New York today? I argue that a culture of fluid urbanism is under way. Water, as a literal and metaphorical principle, is influencing how New York harnesses its maritime pasts to a once neglected, now persistent, reassessment of its water-bound boroughs. New York's growing interest in water ecology and sustainability is here to stay.

Divided into four parts, this book depicts New York in a new phase of landscape redesign, one different from its nineteenth-century provenance. The city is transforming its vision of its future by recalibrating the importance of its shoreline. Flexible approaches to land use, transportation, flows of people, and assumptions about what the city should be are radically rerouting the city's connectivities. New York is slowly embracing its archipelagic geography as an environmentally critical approach to dealing with global climate change and foreseeable storm-water threats.

The book's first part, "Fluid Urbanism," is a phenomenological investigation into the importance of water for New York's self-invention. The fluidity of urban concepts and approaches to city planning are remolding New York's façades. Historic ideas of the maritime city are morphing thinking about changing weather patterns, activating a fluid urbanism between the Hudson River and the Atlantic Ocean. A visual growth plan attentive to low-tech mobility, multileveled, mixed land use, extensive waterfront access, and park spaces for a citizenry seeking flexible leisure spaces is generating unprecedented experiences of the city's extremities.

This book's second part, "Cosmopolitan Frugality," offers cultural readings on the contentious idea of cosmopolitan citizenship articulating New York City. Cosmopolitan urban membership is an ideal embedded in New York's history as an "open" city, a city with a commitment to hospitality that is neither reducible to the idea of the state, nor a reproduction of the older European notion of the hospitable city, the city of refuge.[1] In contradistinction to its European counterparts, New York embodies the ambitious democratic project of perpetually contesting urban belonging, a sense of belonging without a claim to origins, as its emergence lies in a theft of exchange.

Mannahatta was purchased from the Lenape Indians, who did not subscribe to the notion that man could own property indefinitely. According to the historian Robert Shorto, the Lenape viewed their agreement with the Dutch regarding the purchase of Mannahatta as a right to hospitality, with the unspoken assumption that their colonizing "guests" would eventually leave graciously. Hence the exchange of gifts worth sixty guilders for the title deed of purchase of Mannahatta can be considered a theft of exchange on the part of the Dutch, from the Lenape standpoint. Or, from a position offered by Jacques Derrida on the subject of hospitality, one could argue the Dutch overstayed their welcome, and violated their right to visitation, in Indian territory.[2]

Consequently, the impact of globalizing urbanisms on New York's imaginary are multiple. Drawing on the kinetic practices of transnational urban histories such as Second and Third World urbanisms, the Falun Gong, as well as figures such as Mahatma Gandhi, Julius Nyerere, and the Dalai Lama, this second part marks some of the differing philosophies of place-making inflecting local urban practices.

The third part of this book, "Ecological Expressivity," analyzes specific, expressive acts of political cosmopolitan citizenship.[3] It discusses the challenges of sustainable city-making, including the uneven and disruptive arrangements of social understanding. Large-scale interactions have coalesced around the greening of New York's public spaces, drawing attention to the creative citizenship-making mechanisms arising. These performative public enactments further the political and ecological processes of what Jacques Derrida calls "the urban right" to hospitality, a right to an ethics of visitation that was to be negotiated by specifically delineated "limits" within the historic idea of "the city of refuge."[4] The resident, the commuter, the tourist, the stranger, the refugee, and the asylum seeker inadvertently collaborate in this book to test the limits of contentious urban potentiality shaping green urbanism.

The final part of this book, "Maritime Mentalities," consolidates the notion of a fluid urbanism shaped by New York's sea-borne past. It resists the assumption that New York is disassociated from its seventeenth-century oceanic origins as a port city open to the sea. Structured as vignettes, these performative acts, urban accidents, and public rituals investigate a maritime urban sensibility in quest of its democratic fulfillment. The New York in these chapters is partial, incomplete, and distracted—like the city itself. They are windows into a forgotten history when navigation determined the waterfront and the city's shoreline was a gathering place for ocean-faring mentalities. The streets in these chapters bare the raw promise of the right to the city. These streets are sets for a theater of cosmopolitan engagement in the midst of its second act on maritime landscape reuse.[5]

Mnemonic Urbanism

My first encounter with New York elicited an unexpected sense of recognition—a wash of relief after many years of wandering the isolated spaces of American suburbia. A quiet elation filled my person as I inhaled the smells of New York in July of 1992 on Avenue A and Tenth Street, near my apartment on the north side of Tompkins Square Park.

Despite my awareness that the park lay under siege—looking more like riot-torn Brixton, London, during the 1980s than the media-hyped New York of the good life—the aesthetic that arrested me as I walked through the tight-knit sidewalks of the Lower East Side was the feeling of familiarity: this place reminded me of Dar es Salaam, the city I grew up in.[6] Something about the scale and pacing of Alphabet City (New Yorkers' nickname for the then-funky stretch of lettered avenues, A, B, C, and D) resonated with the vernaculars of other deeply ingrained urban memories from other lifetimes and other continents. It is a sensation that has never left me, as I hover by the waterfront on the west side of downtown Manhattan on a cobblestone street called Jane Street, which reminds me more of the narrow streets from the sixteenth-century Dutch colonial era in my parents' home in Cochin (Kochi), India, than it does of American suburban enclaves.[7]

Such disconnected emotional translations of rational urban planning raise the amorphous question of what types of urban sensations are at work in contemporary urban practices. If, from my sensorial encounters with Manhattan's downtown waterfront areas around Pearl Street and Stone Street, the city has more in common with Dar es Salaam, a medieval port city of sixteenth-century fame, than it does with downtown Atlanta, or reminds me more of Fort Cochin and Mattanchery, in Cochin, than it does Los Angeles, then what kinds of connective mechanisms do cities really engender beyond national histories of the urban and colonial cartographies of metropoles?[8]

Furthermore, the realization that most New Yorkers are profoundly provincial—so much so that the prospect of going north of Fourteenth Street in Manhattan, if you live downtown, is a travail—raises the question of what a New York urbanism means. Manhattanites who never go to Brooklyn, Brooklynites who avoid Manhattan like the plague, Staten Islanders who never leave their cars all make the problem of lived cosmopolitan membership a far more local and unstable experience than the macro-vision the word *metropolitan* suggests.[9]

Because Manhattan is an island, it occupies a very particular and distinctive sense of cosmopolitan identity that is more easily shared by other island cities like Hong Kong or port cities such as Cochin, India, and Dar es Salaam, Tanzania, than it is by the kind of suburban spatial imagination of the Bronx or Staten Island, which are communities structured as townships and suburban port-city enclaves, rather than as dense, urban living structured around the water logic of a tightly bound area. This water-defined logic of Manhattan distinguishes the city from the other boroughs, yet Manhattan is also the expression of its five boroughs. It is the connecting, imaginative fascia between Brooklyn, Queens, Staten Island, and the Bronx.

The G Train Phenomenon

New York urbanism is an assemblage of local ecologies. I call it the G Train phenomenon, a term inspired by the train I have to take to get to work. The G train traverses a route between Queens and Brooklyn. Of twenty-seven subway lines, it is the only one that never crosses into Manhattan.

If one lives in Manhattan, which is an island, one has to cross into Brooklyn and switch trains to take the G train deeper into what is increasingly considered the "real" New York, the New York where teachers, nurses, bus drivers, doormen, chefs, hair stylists, waiters, babysitters, artists, therapists, athletes, writers, actors, musicians—just about anyone who isn't a supermodel, stockbroker, trust fund munchkin, corporate executive, or media mogul—are more likely to live. Such a route never entered my imagination before I encountered it, along with my own provincialities. A train route through the city that bypasses Manhattan suggested possibilities of inhabiting the city that I had not fully dwelt upon.

To my insular island logic, all life in New York centers around Manhattan. Yet, there are many New Yorkers who rarely if ever come into Manhattan but actually live in New York all their lives. This unlikely reality surfaced through a course I teach regularly at the Pratt Institute, which is located in

Brooklyn, on the history of downtown Manhattan. Through my students in this rough-edged course, which involves class discussion amid street pandemonium and perpetual road construction, I met born-and-bred New Yorkers from Far Rockaway, Queens, who had never been to historic parks such as Washington Square Park or Tompkins Square Park, both youthful centers of activism. I met New Yorkers who grew up in Sheepshead Bay, Brooklyn, who could count on one hand the times they had ever been to Manhattan, and then only to specific predetermined destinations like museums, and certainly never Greenwich Village or Chinatown. I met New Yorkers who lived on Staten Island, whose idea of Manhattan was the Holland Tunnel. I met New Yorkers who had lived around Washington Square Park, on Bleecker Street, who had never heard of Jane Street, located just two blocks north in the West Village. I met New Yorkers who were born in New York, moved to New Jersey as toddlers, and only returned as adults to the city of their desires and fears. Many New Yorkers had never been to Chinatown or to Bowling Green Park at the tip of the island, where the story of New York begins with the illegitimate sale of the island of Manhattan to the Dutch. New Yorkers avoid obvious tourist locales like Times Square and the Statue of Liberty as much as they do areas outside their daily trudge. These very disparate experiences of New York emphasize that city life isn't just a matter of urban conglomeration but also of local interactions. New York "happens" to its inhabitants.

For many New Yorkers, life in New York City is very local in scale, but also imaginative frameworks brought by migrants. Flushing, Queens, is home to the largest group of Asians. The communities of Stapleton and Clifton, in Staten Island, are home to the biggest population of Liberians outside Liberia.[10] Coney Island, a former Native American settlement and a popular Irish seaside resort in the 1950s, has a considerable population of Pakistanis, Chinese, Bangladeshis, Bukharans, Turks, Kazakhistanis, and Orthodox Jews. Brighton Beach, in Brooklyn, bears the largest contingent of immigrant Russians and is known as Little Odessa. Students from these communities narrate how self-contained these communities' life-worlds are, and how their insularity is shaped against the lure of "the city."

This New York was a city that opened up the question: What does it really mean to live in a metropolis? Is New York urbanism the same as urbanism in other American cities, except on a greater scale? Does life in New York merely produce a more thick-skinned urban dweller with a capacity for denial of creature comforts such as the all-American two-car garage with all its mysterious commodities piled up, or the media room, or closets the size of many New Yorkers' living rooms? Or is there something more to living in an intensely alive city where the real estate is insanely unaffordable, public transportation is heaving under the demands of its daily commuters, and daily chores require a very peculiar local logic that involves a considerable amount of walking? What can we learn from America's first metropolitan city today, and still its largest experiment in urban living?[11]

Contentious Cosmopolitanism

I wrote this book as a memoir of downtown Manhattan at a chaotic time in the city's history, between 2001 and 2013. In this first decade after 9/11, the city's identity was thrown into a maelstrom of emotions. In tension with the rhetorical notions that New York is a city of immigrants, the period after 2001 saw a dramatic rethinking of the city's relationship to its own histories of legitimation.

The New York one popularly encounters is filled with the resonances of Italian, Irish, German, French, Belgian, English, Russian, Polish, Jewish, Hungarian, and African moorings. September 11, 2001, opened up the lesser-known history of New York, which is filled with Arab, Tibetan, Indian, Nepalese, Iranian, Burmese, Pakistani, Bangladeshi, Iraqi, Afghani, Nigerian, Senegalese, Malian, and Lebanese cultural resonances. The surprise that there existed a considerable Egyptian, Syrian, and Lebanese population on Radio Row, a nineteenth-century street that was demolished to develop the World Trade Center, intensifies the importance of comprehending how people of diverse cultural and national origins coexist and collaborate in that ongoing challenge called metropolitan life that envelops most regions of the world today.[12]

The events of 9/11 brought many established New York communities, particularly Arab, South Asian, Sikh, Muslim, brown-skinned—however the categories aligned—under scrutiny and generated tremendous anxiety in the city.[13] Yet at the level of public discourse, the statue of Mahatma Gandhi in Union Square; the statue of Confucius in Chinatown; the images of the Tibetan Goddess of Mercy at street fairs; the wandering bands of Hare Krishna devotees in Washington Square Park, Tompkins Square Park, and the Union Square subway station; mosques in the East Village, Brooklyn, and Queens; devout Muslims genuflecting toward Mecca by gas stations at midday; and the images of the Dalai Lama displayed by the Tibetan community in prominent thoroughfares in New York City every year all signal how cosmopolitan life in New York has always incorporated philosophies of the city that are international and intimate, in ways that only walking cities can achieve.

The aura of catastrophe that overtook the city following that harrowing September day in 2001 confused perceptions of what New York cosmopolitanism is, causing a retreat from the radically democratic and unstable spaces of public encounter, spaces for which New York is most famous, a retreat toward a culture of rescue, salvage, mourning, and recovery. A distinctly anticosmopolitan sentiment crept into the city. For many South Asian cabbies, particularly those of Sikh and Muslim cultural moorings, the anxiousness to demonstrate their allegiance to New York City was displayed through the conspicuous posting of flags and the discarding of the turban—public signs of solidarity with New York City and a cautious reassessment of their own relationship to New York urban space.[14]

A Poetics of Hospitality

Now faced with an ever more ruthless logic of the market's determining life in New York City, the city tenaciously clings to its fading aura as a city of hospitality whose doors have historically been open to those who seek refuge.[15] It persists in asserting the elusive, forced idea of cosmopolitan citizenship. As the financial shocks and global downturns ripple through the boroughs, people continue to imagine, dream, desire, protest, love, and lose themselves in the city.

This book takes as its premise a fact of productive tension: New York City is a metropolis whose singular sense of itself is its identity as a distinctive water-bound world city, while also comprising a regional network of urban clusters that shape its global visage. New York's cosmopolitan identity is polyvocal and always in the process of unraveling. Its spectacular guises force the dwellers of its boroughs to ask: What is this cosmopolitan perception? How does one begin to articulate cosmopolitan citizenship from within a city perpetually in reconstruction?

The following pages are an offspring of those taut, uneasy expressions of urban belonging. They form a meditation on the public sentiments and social feelings that produce urban subjects as citizens of the polis. Written as street scenes capturing fluid urbanism, this book is less E. B. White's idea of patrician New York and more immediately the New York of a Gambian or Pakistani cabbie who has traversed the entrails of the city for two decades and pushing. These explorations unfold against a backdrop of New York's larger-than-life persona as an urban entity, where fiction, big money, brutal real estate, and hard-nosed speculators compete with the grit and intimacy of real lives lived hard for reasons forgotten elsewhere, or not yet begun to be realized.

I offer a reportage of a world city at a critical point in its vast and undulating trajectories between the mayorships of Rudolph Giuliani and Michael Bloomberg—the period leading up to 9/11 and the aftermath of recovery; the time when the city convulsed through the dot-com bust and the 2008 financial meltdown, with the bottom falling out of New York's way of life; the time of the $700 billion bailout package and Wall Street's implosion. It is a time when New York's grim 1970s threatened a return: shuttered stores, bankrupt businesses, abandoned storefronts, expensive furniture discarded on street corners, favorite bookstores, cafes, and hardware stores gone. Many New Yorkers without jobs are still trying to retool and reinvent a life after 2008.

It is also a time when Barack Obama became the first African American president, and New Yorkers stopped living in a state of perpetual Orange Alert. Two dramatic plane episodes on the Hudson River have led to a rethinking of chaotic airspace management over the Hudson River, which sees over 25,000 unmonitored helicopter rides a year. The year 2009 saw the blocking-off of traffic from Times Square for the first time, the reopening of the Washington Square Park fountain aligned with the Washington Square Arch, and the 400th anniversary of the "discovery of Mannatus" by Henry Hudson on September 6, 2009, commemorated by the arrival of eleven Dutch barges from the Netherlands, floating down the Hudson River.

In 2010, the timely detection of a car bomb set to explode in Times Square threw the city into a momentary shudder. But despite the resulting evacuation and unease, the incident did not deter people from enjoying the hospitality of a balmy New York night.

Cosmopolitan New York is a culture of risk its citizens are fully in tune with. The ecological catastrophe of wastewater contamination along the Hudson River in the summer of 2011 emphasized how regionally interconnected New York's environment is. The polluting of the river was a disaster for cities along its banks. It was a wake-up call urging the responsible use of the city's irreplaceable resources, such as the Hudson River, which requires a persistent commitment to sustainable water and waste management.

Ecological risks entailed by metropolitan living also include expansive pleasurable moments, such as the city's expanded greenways interweaving its five boroughs. People can bike from Manhattan to Brooklyn, Queens, or the Bronx, or pedal from downtown Manhattan across the George Washington Bridge. It is a risk built on a shared understanding that New York's cosmopolitan approach to hospitality is a fragile utopian undertaking, as the Occupy Wall Street protests in Manhattan's financial district display. Uneven outcomes of cosmopolitan citizenship threaten New York's promise as a city of hope, of arrival, of being a doorway to cosmopolitan becoming.

New York performs the idea of itself passionately, and resilience is its most impressive feature.

fluid urbanism

New York is reimagining itself environmentally. The city has awakened to a belated, ecological vision of its futurity. It is an era of fluid urbanism. Water has become prominent in the city's development. In tension with earlier concepts of the city that sought to tame, cultivate, and bury nature under macadam within the sanitary modern city, the emerging ecologically conscious city struggles to dialogue with nature. Its plazas incorporate nature into their atriums. Its façades increasingly reflect the moods of nature. Its walls seek to draw upon the heat of nature. Its populations seek direct encounters with nature in their immediate neighborhoods.

A nuanced understanding of the built environment has been transformed by a renewed interpretation of New York's context: that it is a city surrounded by water. This aqueous urban vision seeks to learn from other cities, from other countries. It is an urbanism forged amid a new malleability in the state of things economic, environmental, and conceptual. The city is at a crossroads. Infrastructure is being forced to painfully accommodate a growing desire for an energy-efficient metropolis. A fluid perception is transforming how people occupy space in the city.

This section marks New York's cultural shift away from earlier historic periods, including the era of Robert Moses, when the waterfront was a blighted region, divorced from the civic life of the city, to a contemporary moment where multiple voices, ideas, interests, and planning objectives compete to find a more sustainable outcome for the city.

This is an era when the fluidity Jane Jacobs, the writer and urban activist, advocated in her account of the balletic dance of the street has expanded to include a range of alternative vehicles, such as toddler strollers, Segways, foot scooters, and cycles, alternatives demanding new approaches to traffic flows. Notions of what the neighborhood is have completely changed since Jane Jacobs's time, the 1950s. There are many more eyes on the street today than she probably would have liked, and many of them are technological as well as demographically different from Jacobs's cozy idea of the neighborhood butcher and store owner. Yet, many of Jacobs's concepts are still relevant to the flux and pace of the streets.

More flexibility is sought by communities imagining their neighborhoods, a notion first advocated by Jacobs in her concept of mixed-use approaches to development. New York's building boom is necessitating a rapid-fire reeducation of planners, architects, civic authorities, activists, commuters, and neighborhood eyes on the street. Their insights are part of the larger process of fluid urbanism, in which the condition of cosmopolitanism plays a defining role. Cosmopolitanism is a critical component of the larger urban process known as metropolitanism.

■ Metropolitanism is the fastest growing urban experience today. The rise of megacities across Asia, Latin America, and Africa, alongside the sprawling cities of the geographic north, means that greater populations are experiencing the particular challenges of metropolitan living. Dense living within metropolitan regions involves particular relationships of place-making. This process is as varied as the cities that are exploding around the world. Desert cities, mountain cities, port cities, island cities, each presents its own set of ecological parameters through which the notion of an urban experience is being imagined.

New York City is the first prototype of dense urban space that has self-consciously documented, historicized, mapped, and perpetuated notions of a distinctively local set of metropolitan experiences. To this extent, the city continues to offer lessons on how urban space can be generated, reinvented, recuperated, and managed. At the heart of this urban infrastructure called the metropolis is the idea that large cities like New York are interwoven into their regional geographies. As a metropolis, New York City draws into its vortex the multitudes.[1] Who these multitudes are is the ongoing focus of metropolitan thinking.

Metropolitanism is a state of engagement. It involves governmental, infrastructural, and nonquantifiable human impact shaping macroplanning decisions within urban networks. The language of architectural and urban planning frequently contains abstract notions of human interaction within metropolitan regions. Visual images of people inserted into projected urban plans and imagined social activities inscribed onto invasive architectural projects attempt to incorporate human expressivity into metropolitan visions. In practice, people are frequently evacuated of locality, variety, and particularity within planning, as the larger, all-encompassing notion of a metropolitan region imposes the aura of anonymity and enormity on cities such as New York.

From the point of view of the pedestrian, living in a large city involves multiple layers of urban experiences. Some are local, others are global and generic, familiar to any big city. New York City offers particular lessons on the abstract notion of metropolitanism. For one, New York metropolitanism is an experience of scale. Demographically the most culturally diverse city in the world, where peoples of every possible nationality have citizenship rights to being a New Yorker, the challenges of studying New York demand a fluid approach. That approach involves broadening the categories of city-making beyond the infrastructural logics of planning to include human embodiment.

Metropolitanism is a condition of life as well as an approach to city planning and governance. It necessitates a fluidity of encounter, of negotiations between civic authorities and cultural entities. More pointedly, in New York one experiences a walking metropolitanism, and, to that effect, it is grounded in the local. The city has developed textures, inflexions, atavisms, philosophies, monuments, performances, festivals, and urban rituals to reaffirm the social experiment understood as cosmopolitan urban citizenship.

However, metropolitanism is not cosmopolitanism by another name. Metropolitanism encompasses regional processes and networked transportational realities producing spatially determined urban subjects. It is a macroscale urban experience that includes intercity dependency and citywide subject-making processes. The metropolitan experience does not by default nurture cosmopolitan citizens, though cosmopolitanism is implied when invoking dense urbanity. The degree of cosmopolitan potentiality differs from one metropolis to another, and depends on the strength of civil society and legal frameworks.

■ Cosmopolitanism, an aspect of metropolitan life, is a fraught and contingent ideal historically embedded in the very idea of the polis, or city. It contains the idea of world citizenship, a sense of belonging that exceeds the finite definitions of human association, such as nation, race, and language. The term does not imply movement outside the city.

Life in the metropolis necessitates the cultivation of political cosmopolitanism. Political cosmopolitanism is the legal production of urban coexistence through dialogue, debate, consensus-forming, and voting practices engaged by the public. It is the uneasy inculcation of a particular type of citizen-subject: one who must dwell alongside peoples of contrasting cultures, with frequently conflicting understandings of how to live in cities. This citizen-subject, the political cosmopolitan, can be an individual who may not have traveled outside the city, but whose notion of a city is deeply influenced by the quotidian practices of sharing public spaces with its citizenry. Big cities like New York reveal peculiarly entrenched, local characters whose larger-than-life lived experience within the city is more often than not accompanied by a provincialism incomprehensible to better-traveled non-New Yorkers. Such a subject is shaped by social and legal precepts in a culture of mutual tolerance, of cosmopolitan affiliation anchored by binding notions of public right. This state of perpetually negotiating cosmopolitan perception of what the city is and should be is the process of fluid urbanism.

The inherent tensions between a regional metropolitanism and a pedestrian-centered political cosmopolitanism intertwine to produce New York's particular manifestation of fluid urbanism. For New York, an archipelago of islands connected by bridges and tunnels, water is a defining, though till recently much-neglected, topography shaping its urban identity.

Fluid Urbanism foregrounds the literal fluidity of New York's location on the Atlantic Ocean, the presence of rivers, buried water sources, and a world-class harbor that was once its maritime claim to fame. The term's conceptual fluidity captures peoples of disparate languages and cosmologies who formally cultivate a shared living space with delineated boundaries recognized as those of the cosmopolitan city.

CHAPTER 1

water ecology, island city

In a map of Manhattan published in 1865, lovingly etched with fine lines and exacting details, the viewer discovers a scenario of a sheltered island filled with rivulets, forests, bays, swamps, meadows, marshes, hills, ponds, reclaimed land, and sewer lines (see figure 1.1). The island is narrow but lush with rivers and estuaries. The land mass brims with a hilly topography of trees, cliffs, and coves. The viewer is reminded of a gigantic rock covered with oyster shells that lie glistening in the clear sun of an island that marks the beginning of an enormous continent extending westward.[1]

Colonel Viele's marvelous map of a Manhattan that no longer existed by the time he was scaling its details in 1865 is a glimpse of what the island of Manhattan meant ecologically, as a marine biosphere. It depicts a landscape teeming with a variety of birds, fish, and fauna and flora, from the island's narrow, low-lying tip in the south to its northern rocky mass. By the late twentieth century, the imagined space between the idyllic and the lived would spawn the densest artificial organism in the urban world. A symbiosis of water ecology and island city, New York City continues to be studied as a contrived biosphere designed to foment some of the most challenging human experiences ever imagined. It continues to pose hard questions about the potentialities of democratic coexistence amid crowded cosmopolitan living.[2]

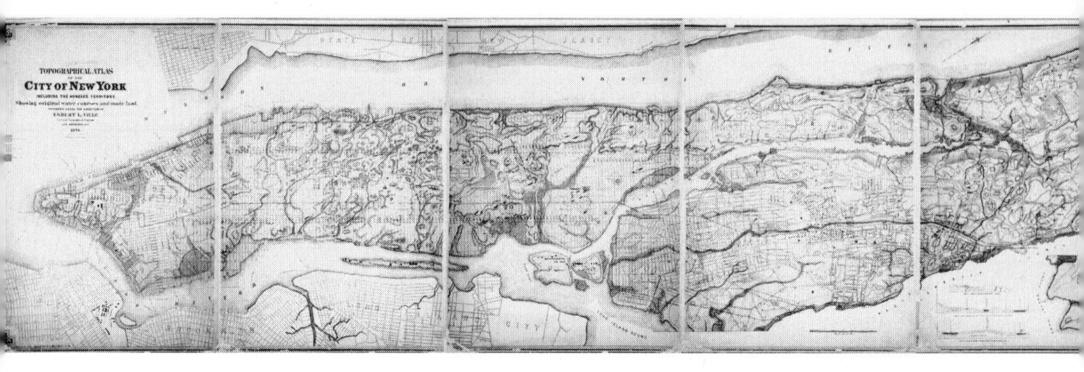

FIGURE 1.1 *Sanitary and Topographical Map of the City and Island of New York Prepared for the Council of Hygiene and Public Health of the Citizens Association*. Produced under the direction of Egbert L. Viele, topographical engineer, 1865. Courtesy of the New York Public Library.

Water Ecology

Manhattan is an accident of water.[3] Its locality as an island determines its water-bound ecology. The borough is surrounded by three bodies of water: the Hudson River, the East River, and the Harlem River. Of these, the Hudson River is the most famous. Its claim to distinction is its historic vantage as the gateway to the Erie Canal and the vast regions of the American hinterland through the Great Lakes. Less understood is the Hudson's unique ecology as a tidal estuary.[4]

The Hudson River originates in the Adirondacks and flows south toward the New York Bight in the North Atlantic Ocean. A tributary, the Mohawk River, meets the Hudson along its journey downstream. South of Manhattan, the North River, as the Hudson is referred to by local mariners, fuses with the eddies of the Long Island Sound, flowing from the northeast through the East River tidal estuary toward Upper New York Bay. West of the Hudson River, the waters of Newark Bay flow into Upper New York Bay, adding to the churn of New York Harbor. Coursing through the Verrazano Narrows, Upper New York Bay opens into Lower New York Bay, merging with flows from the Raritan River and Jamaica Bay, finally pouring into the North Atlantic.

Known to the Lenape Indians as "the river that flows both ways," the Hudson's tidal action propels the flow of water in both directions. According to the landscape ecologist Eric W. Sanderson, at one point in its evolution the Hudson River experienced four tides a day.[5] Surging all the way to

Troy and streaming back to the Atlantic Ocean past Manhattan, the Hudson's confluence with the Long Island Sound and the Atlantic Ocean activates currents, flows, waves, and shifts in history and culture, creating the transforming habitat of the island of Mannahatta, today's New York City.[6]

Both fresh water and saltwater flow around New York City, impacting what Sanderson calls ecological neighborhoods.[7] Abundant with rocky streams, lakes, ponds, and reservoirs, many of Manhattan's water traces are now buried underground. By the nineteenth century, Minetta Brook, which courses southwesterly from Washington Square to the Hudson River, and ponds like Collect Pond were built over. The island's canals and slips, of which those marked by Canal Street and Broad Street are the most famous, have been paved, and its low-lying swamps and marshes have all been filled in, creating downtown areas such as Alphabet City, South Street Seaport, Water Street, and Greenwich Village. Times Square, in Midtown, was once a red maple swamp.[8] Central Park was formerly a swamp that was reclaimed by Manhattan's first African Americans before it was transformed into the lungs of New York. The gaping, exposed "slurry wall" in the cement bathtub of the ruined World Trade Center's foundations, with its metal pegs, is the single most powerful reminder of the artificiality of the city's expanded land mass, hooked into the bedrock of the rocky island's Manhattan schist with pilings over 150 feet long.[9]

Island City

The environmental and urban catastrophe of 9/11 accentuated certain structures of metropolitan perception essential to New York City and particularly Manhattan. Ecology emerged as a crucial aspect of life on the island. The issues of air and water quality became critical sources of anxiety in the hours following the disaster.

Manhattan's water-defined boundaries made it impossible for those living downtown, near the World Trade Center, especially those who were dependent on the public transportation system, to leave the island immediately following the catastrophe. The suspension, due to successive bomb threats, of trains out of Grand Central Station, and the systematic cordoning-off of the perimeter south of Twenty-Third Street, and later Fourteenth Street, and subsequently Chambers Street and the downtown area, from vehicular traffic over a period of a few weeks, raised pressing questions of sustainability, locality, and geography, and highlighted the importance of water-going public transportation networks that are normally subsumed under the enormity of the bridge and tunnel networks of New York City. Ferries that customarily transported people across the Hudson River to Staten Island and New Jersey were disrupted. Enveloped in acrid clouds of black dust and unable to leave downtown Manhattan easily, given the shutdown of all egress from the island, one felt acutely the tenuous reality of living on a confined, manmade island.

This visceral metropolitan sensation of environmental vulnerability was accompanied by a profound distrust in the Environmental Protection Agency's unconvincing assurances that the air quality around Ground Zero was adequately safe for human occupation, even as people in the vicinity started developing severe respiratory diseases, particularly infants, asthmatics, rescue workers, and the elderly. The poor air quality for weeks following the environmental disaster, despite the agency's pronouncements to the contrary, accentuated a dawning realization of how precipitously located Manhattan is as an urban construct.[10]

The unexpected calamity has since forced a rethinking and new appreciation for the island's ecological composition.[11] The impending scenario of swelling oceans and the rising waterline of the seas is tangibly felt on the island as inhabitants regard the sinking land around Water Street and Front Street on the island's tip, where its oldest and most historic buildings stand (see map 1).[12] Water Street was underwater when Henry Hudson glimpsed the island in the early seventeenth century and is an ongoing reminder of the delicate balance between bedrock, landfill, tidal marshland, seagrass beds, rocky streams, shifting oceanic patterns, and built environments.[13]

The cataclysm brought to the fore the challenges that a cosmopolitan ethos entails. A majority of those who died in the World Trade Center disaster did not live in Manhattan or even in New York City, but rather commuted into New York for work. This reality raised questions about how a metropolitan identity actually works emotionally and physically as a social technique of urban imagination.[14] What makes New York at once distinctive and increasingly generic? This flexible labor force of commuters, tourists, travelers, and workers compounds the difficulties of understanding New York's cosmopolitan composition. Its subsequent need to mourn and honor its dead in a manner that adequately acknowledges the scale and breadth of the individual identities concerned encompasses a regional notion of metropolitan identity, at once of the city but not from the city.

With one of its signature, though controversial, landmarks destroyed, the idea of New York City as a visual, psychological, physical, mental, and

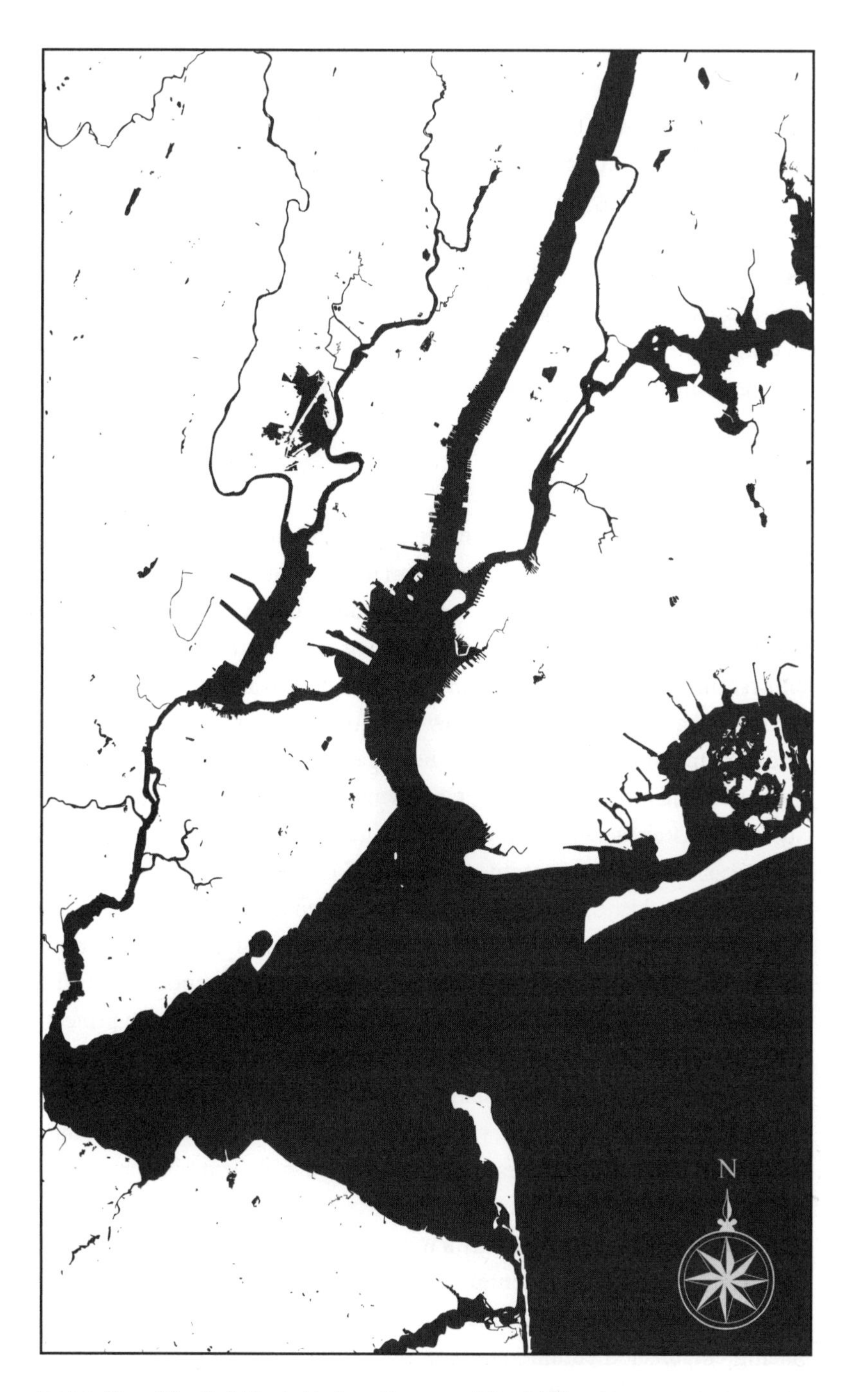

MAP 1 *Map of New York City Archipelago*. Courtesy of Daniel Hetteix.

organizational entity has undergone an intense and city-transfiguring makeover.[15] From the ashes of the destruction, a greater awareness for the island's water-bound topography was forced upon the city. By extension, the city's waterfront neighborhoods and their more ambivalent developers are taking the opportunity to embrace a more water-centered, albeit privilege-determined, relationship with New York's waterfront in a post-industrial envisioning of the New York waterscape.[16] In a shift away from Robert Moses's conception of a New York that severed its inhabitants from its waterfront, the New York of Mayor Michael Bloomberg seeks to establish a growing relationship between its inhabitants and its waterfronts.[17]

Reclaiming New York's Shoreline

Alongside New York's efforts to move beyond the narratives of mourning and grief, the city's commitment to clean up the tidal marshes through the slow reclamation of its tidal estuaries is transitioning its waterfront into a place for meditation, public leisure, urban rejuvenation, and sport. For the first time in recent years, New York's archipelago-city ecology is finally becoming an integrated part of how the city imagines its future. This extended reimagining has not always been people-centered, but it has been consistent and sweeping. Rising oceans and attendant concerns for security and disaster management are not an inconsequential part of this new awareness of New York's status as an island city whose bridges, tunnels, waterways, roadways, and airways present the densest nexus of transportation economies shaped by water flows and geography in the United States.

Across the boroughs, large public and private developments, such as the Brooklyn Promenade, the Hudson River Park, the High Line Park project, the Domino Sugar Factory, the Gowanus Canal redevelopment plans, the transformation of Governors Island, the East River developments, the Red Hook revitalization initiatives, the Atlantic Yards developments, the Coney Island Redevelopment debates, and the projected visualizations for a future South Street Seaport are dramatically altering fragile social arrangements between populations and the city's waterfront boundaries. Some of these transformations are detrimental to the neighborhoods in which they are located. Using the laws of eminent domain, developers are ruthlessly displacing established commercial and residential neighborhoods, despite wide-ranging protests and neighborhood interventions. Elsewhere, less volatile expansions are making new lifestyles possible for waterfront areas.

Cumulatively, New York's coastline is in a historic transition. The city has opened up its beaches and shoreline as places of leisure. Water activities, such as sailing, ferry boat rides, surfing, and free community kayaking, and infrastructure such as restful wooden boardwalks, pedestrian coves, benches along the riverfront, and wireless networks have expanded the role of the city's shores within its imagination.

New Yorkers can now swim around the island every year, from August to October, in "swimathons" organized by Swim New York City. The annual swimathons along New York's shorelines have visually redefined the quality of New York's waters for its inhabitants. These events, including the Brooklyn Bridge Swim, the Ederle Swim, the Little Red Lighthouse Swim, the Lady Liberty Swim, and the Governors Island Swim, invite physical immersion in New York's rivers as an annual calibration of its waters' safety. The annual Manhattan Island Marathon Swim New York—28.5 miles around the island of Manhattan, in a counterclockwise circumnavigation of the city—is a spectacular culmination of this slow reclamation of the waters around New York City (see map 2).

Every May, designated Bike New York Month, cyclists can participate in a number of annual "bikeathons" within the city—the Five Boro Bike Tour, a forty-two-mile bike ride beginning in Battery Park, Manhattan, and culminating in a Staten Island Ferry ride back to Battery Park is the bikeathon to end all bikeathons. With 32,000 cyclists registered in 2012, the Five Boro Bike Tour traverses all five boroughs, carving uptown through the center of Manhattan, Harlem, and Queens, to Brooklyn and the Verrazano Narrows Bridge, to Staten Island's Ferry terminal and back to Manhattan's Battery Park. Not to be sidelined, runners have expanded their claims to New York streets with the mini-marathons that culminate in the early November grand finale, the New York City Marathon, which in 2011 had 47,000 runners coursing through the city's arteries.

An imaginative rethinking of where the city should go visually, after the collapse of the vertiginous heights of the World Trade Center Towers, opened up environmentally conscious ways of talking about what buildings can, and should, do. Initial planning discussions among the city's residents and officials explored the opportunities and constraints of reimagining downtown Manhattan on a scale never before attempted. Considering utopian desires for a public bird sanctuary at the former footprint of the World Trade Center site and calls to rebuild exact replicas of the two towers, the city's planners, activists, citizens, and bereaved families de-

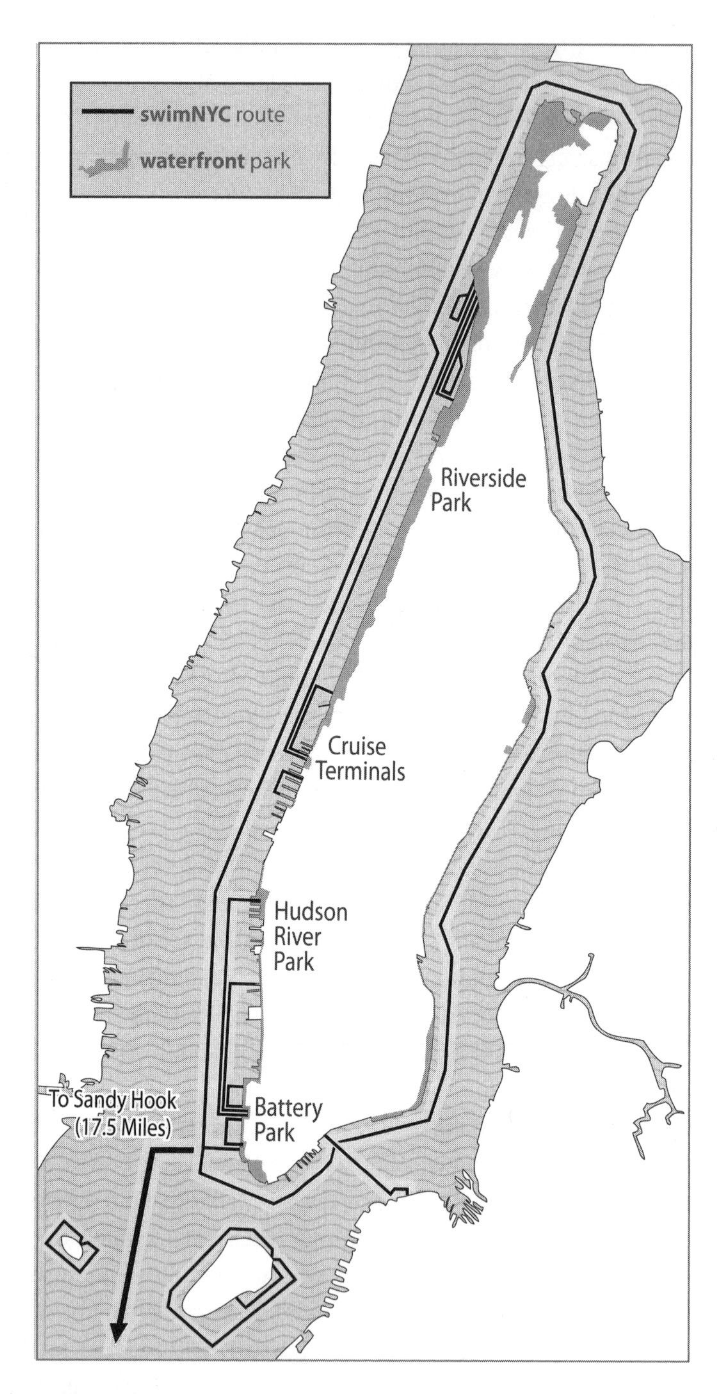

MAP 2 *Map of New York City Greenspace around Waterfront, and Swim NYC Routes.* Courtesy of Daniel Hetteix.

bated numerous designs. A central concern was how to transform a now-toxic graveyard of prime real estate by the water into a place of the future, rather than just a cemetery of the past.

After many architectural renderings and years of planning debates, the commercial imperatives have driven the final plans. The idealism, architectural virtuosity, and originality of the initial proposals for the site proposed by the visionary architect Daniel Libeskind have been scaled back in favor of the vertical, bunker-shaped, commercially determined One World Trade Center, formerly known as Freedom Tower. In the process, however, the necessity to rethink mixed-use neighborhood revitalization, commercial viability, and the ecological and environmental durability of New York City as an island city of bridges, tunnels, bike paths, pedestrian walkways, and ferry systems connecting five different land masses, all surrounded by water, has led to a resurgence of interest in the marine ecology of New York City.[18]

Subsequent years of discussions have fortunately opened up the importance of New York's local and regional habitats of marshes, swamps, and forests as a delicate biosphere of great diversity that needs to be preserved, protected, and cultivated as part of the larger scenario of the ecological viability of a global metropolis.[19] Environmental planning initiatives since 2001 have sharpened the city's demand for energy-smart buildings using green design and environmentally sustainable technologies. An influential contribution in this direction is the High Line, on the West Side of Manhattan, a collaboration between James Corner Field Operations (Project Lead), Diller Scofidio + Renfro, and the planting designer Piet Oudolf. Their innovative undertaking, a public park project emphasizing adaptive reuse, has proven a provocative leap to further dialogue on creating green industry alongside sustainable design.

The High Line is an invitation to reimagine how New York's greening initiative can utilize its existing heavy industrial infrastructure to realize its future park and recreational needs. It has not, however, been an initiator of green industry, which some critics argue should be incorporated into a vision for greening New York City. Merely creating a green city of consumption and leisure is an untenable goal. It squeezes out lower-income communities and raises rents in formerly affordable buildings.

A sustainable plan for greening a metropolis has to be accompanied by comprehensive plans for production and job creation in the green sector of city growth. Eco-efficiency needs to be accompanied by a feasible projection cost to maintain parks, waterfronts, streets, and public spaces. These hidden expenses make greening initiatives appear a luxury when the economy is slow and unemployment rife in the city. Consequently, a balance has to be woven into the city's long-term, sustainable planning commitment, so that greening initiatives include job creation and access to all income demographics.

Between Preservation and Growth

The post-9/11 receptivity to reimagining the diminished skyline has led to the swift rise of signature buildings, sometimes out of context with the neighborhood, buildings designed by Frank Gehry, Santiago Calatrava, Jean Nouvel, Rem Koolhaas, Tadao Ando, and Richard Meier. In tension with the measured and thought-out integration of heavy industrial infrastructure, as realized by the High Line project, is the more opportunistic and chaotic development along the Bowery, the Far West Village, South Village, and the Meatpacking District. The growth in environmentally conscious engagement with land use has been predictably accompanied by a new onslaught against zoning laws.

The Greenwich Village Society for Historic Preservation (GVSHP) put forth a proposal in 2006 to designate the South Village a historic district. The GVSHP's efforts to rezone the South Village, the Far West Village, and Hudson Square demonstrate the conflict between long-standing protections for low-lying industrial areas and developers' impetus to acquire variances, developers who often demolish historic façades. The tension between preservationists who wish to protect the fast-disappearing character of New York's historic neighborhoods, and the Bloomberg administration's rezoning of New York in favor of upscale real-estate investments has created a battleground of interests across the city. On the one hand, there is pressing need for more affordable housing in New York. On the other hand, there is an urgent need to protect low-lying industrial neighborhoods that have not been designated as heritage sites.

This fractious lag between preservation laws and the rapacious hands of developers has caused an ongoing battle for the future of New York's skyline. It is the implied tension of a metropolis: to preserve the historic past while continuing to build densely toward the skies, to fulfill the cosmopolitan demands of the city. These conflicting interests play their scripts to great cacophonous protests around New York. Atlantic Yards, Hell's Kitchen, the Bowery, the Far West Village are some of the recent scenes of contestations.

Under the Bloomberg administration, the sentiment on the ground has been that housing growth has been made in the interests of the wealthy and the middle class. The pretext of affordable housing has accompanied much of the fancy high rises transforming formerly industrial and residential neighborhoods in Manhattan and Brooklyn.

Perhaps the most striking example of the accelerating threat to Manhattan's remaining industrial neighborhoods is the poignant destruction of historic neighborhoods along the Bowery, a region of the Lower East Side whose mix of low-lying wholesale, commercial, and flophouse activities cultivated a bohemianism legendary in much of New York's cultural life. The area's distinct flavor of cut-rate commercial lighting stores, wholesale restaurant suppliers, flophouses, and bars allowed a distinctly New York cultural milieu of artist loft spaces, alternative theater spaces, second-hand furniture stores, underground music clubs, and small commercial enterprises to coexist amid the tangible dereliction around the neighborhood.

Today, an endearing and irreplaceable area of old New York has been erased. At the junction of Houston Street and the Bowery, industrial warehouses have been demolished and dramatically restructured for a dense, suburban-scale urban lifestyle. The obtrusive, out-of-scale incursions of residential developments that have caused rents to soar in such formerly industrial areas as the SoHo district and the Bowery is a sign of what is to come. These areas have not been adequately demarcated as historic districts protected by the New York City Landmarks Preservation Commission and the New York City Department of Planning. Consequently, historic preservationists, ecologically minded architects, and environmentalists alike view these vulnerable industrial zones as the last outposts of New York's air rights, to be protected with great effort and at any cost.[20]

The construction of Trump SoHo Hotel demonstrated that more vigilance is needed from neighborhood, municipal, and civic authorities who serve as watchdogs of public interests, and not corporate and private interests. Bulky, tall buildings that block light and air in low-lying neighborhoods detract from a locality's quality of life. Furthermore, the influx of transient wealthy denizens that places such as the SoHo Trump attracts often congest low-impact neighborhoods with heavy tourist traffic without giving anything back to the neighborhood.

These dramatic changes since 2001 have raised responses from the street, about how people of the city are being affected by the massive reimagining of New York City. Many New Yorkers are once again reminded of the top-down planning reminiscent of the heyday of Robert Moses's grand schemes for New York. In reaction, efforts are being made on the ground to rally different neighborhoods, activists, and planning groups to force a more people-centered perspective of the future of New York.[21]

The continuing question in this ongoing surgical incursion into the city's neighborhoods is the future of cosmopolitan citizenship. How will the New York of hardworking middle-class families and the economically disenfranchised survive this new visionary leap into a revitalizing New York? Where does New York's culture of small businesses, mom and pop stores, low- and middle-income families, and the invisible populations of its new immigrant communities fit into the rapidly gentrifying landscapes of its future?

A disturbing response to this query lies in the instance of the developer Bruce Ratner's much-opposed $4.2 billion Atlantic Yards project, near downtown Brooklyn, a project that deployed eminent domain to propose a mixed-use development that affected nine property owners and a dozen residential and commercial buildings in an established Brooklyn neighborhood. Despite vociferous protests, public outcry, and wide discontent at Ratner's disruptive plan, which includes an arena for the now Brooklyn Nets basketball team, the project was approved by New York's appellate court in May 2009. The court ruled that the deployment of eminent domain to acquire private property to construct a sports arena does not infract the state constitution.[22] In the face of such ruthless real estate development, how do people remain involved in this ever-expanding city of the skies? What kinds of citizenship-making practices does New York have to offer at this point in its history?

CHAPTER 2

transocеanic new york, city of rivers

September 2009 was the four hundredth anniversary of the sighting of Manhattan by the British expeditionist and employee of the Dutch East India Company, Henry Hudson.[1] Marking the historical date of September 6, 1609, the Quadricentennial spurred great commemorative interest on the part of the Dutch government, as well as New York State. Given that past centennials did not generate such a concerted effort to reclaim a lost Dutch past, the populist evocations of Dutch influence in the founding of New York appeared more marketing strategy than historic homage. Nonetheless, a language of Dutch cultural reclamation percolated through the New York landscape and Hudson River Valley.[2]

The festivities reinforced New York's economic prominence, while shedding light on its Dutch cultural roots. Thus 2009 became a year of revisiting and recalibrating what Manhattan is, what New York has become, and how the history of the city's past four hundred years now comes to haunt its future in the face of a collapsed world economy, global climate change, impending water scarcity, and rising oceans.

At the beginning of the apocryphal narrative of Mannahatta's discovery by Henry Hudson is the story of a dream: the dream of a passage to Asia through the North Pole, a passage the Flemish cartographer Gemma Frisius believed sliced across the Arctic. This legendary strait appeared on Gerardus Mercator's map of 1569, labeled "arctic strait."[3]

Gripped by this possibility seventeen years prior to Hudson's "discovery," in 1609, the Dutch geographer Petrus Plancius created an influential map in 1592, called Nova et exacta terrarum tabula geographica ac hydrographica, indicating the possible existence of a route to Asia through the North Pole.[4]

Seized by this fantasy, Henry Hudson embarked on a mission to find a new route to Asia for the Dutch East India Company; but, instead, he discovered a great river filled with oysters and beavers: "The sixth in the morning was faire weather . . . The Lands they told us were as pleasant with Grasse and Flowers, and goodly Trees, as ever they had seene, and very sweet smells came from them."[5]

By 1592 Plancius had marked a number of now-forgotten Asian and African ports of great strategic relevance to the Dutch empire at the time. Cochin, Cape Comorin, Calicut, Goa, Zanzibar, Malindi, and Mombasa, to name a few points of Dutch interest, are all clearly delineated on a map whose land masses have vast unmarked terrains at their centers, terrains suggested by decorative cartouches. Plancius's map was made before the discovery of Mannahatta. The map supports the reason the Dutch East India Company founded its conglomerate of Dutch interests abroad: to find a way to control Asian trade.[6] Plancius's map marks the colonial ports in which the Dutch were invested for their colonial enterprise.

The emphasis on bodies of water in the emerging science of cartography during the early sixteenth century accentuates the oceanic imaginaries of navigational maps. Port cities critical to the period's global trade are marked in great detail in these early navigational charts. Many of the important maritime cities in the heady days of early Flemish and Dutch cartography are obscure in the twenty-first century. Gerardus Mercator, in his Nova et aucta orbis terrae descriptio ad usum navigatium emendate (World map for use by navigators) (1569) assigns significance to the port cities of Cochin, Quilon, and Zanzibar. The Dutch were very invested in oceanic networks between Asia and Africa, connecting Deshima to Ambon to Cochin and Kaapstad (Cape Town) during the period leading to Mannahatta's colonization and the relinquishing of New Amsterdam.

In an elaborate map framed by the signs of the zodiac; the four seasons; the four elements of wind, fire, water, and air; and the seven wonders of the ancient world, the hydrographers Joshua van den Ende and Willem Blaeu created the fabled Nova totius terrarum orbis geographica ac hydrographica tabula (New geographic and hydrographic map of the entire world) (c. 1630). Bodies of water are the focus in this extraordinary dreamscape: Mare Pacificum, Mar de India, Mar del Nort (North Sea), Mare Atlan, and Oceanus Ethiope are identified. The eye wanders across the hydrographic mapping of oceanic and river flows in the rendering. The map by Ende and Blaeu proposes a new way of comprehending the world. It is a world structured by water interspersed by land.[7]

Another rendering of oceanic knowledge in the emerging geographies of the time is Joan Blaeu's Asiae descripto novissima (The most recent description of Asia) of 1659. By the time Joan Blaeu's map emerged, additional detail appeared in the interconnected correlation between bodies of water: Mare Arabicum et Indicum, Oceanus Chinen, Mare Rubrum, Mare de Mecca et B'Omar Corsum, Sinus Arabicus, Sinus Gangeticus, Gosjo de Bengasa, Oceanus Tartaricus. These expanding waterways open up the navigational networks across maritime trade routes, explicitly connecting North America to the Atlantic Ocean, Indian Ocean, Arabian Gulf, Arabian Sea, and South China Seas.[8]

Sea Charts of Memory

The noted hydrographer Lucas Jansz Waghenaer, in his Spieghel der zeervaert (1584), revolutionized the mapping of coastlines in detail. His influential sea charts were published in English as the Mariner's Mirrour (1588) and in Latin as Speculum nauticum (1591). Using a sounding line, Waghenaer paved the way for the precise measuring of ocean depths and detailed nautical charts.

Deploying Waghenaer's technique of soundings in their fourth journey across the Atlantic in search of the passage to the riches of the East, Henry Hudson's crew charted the coastlines, sea routes, and shore soundings of their voyage in 1610. These sea charts, titled *Tabula nautica*, were published by Hessel Gerritsz in his Descriptio ac delineatio geographica detectionis freti sive transitus ad occasum supra terras Americanas in Chinam (Description and map of the discovery of an inlet or passage to the West above the American lands to China) (1612). The *Tabula nautica* remains the only detailed record of Hudson's fourth and last voyage to North America. These interpretive maps resituate the narrative of the "discovery" of Manhattan within a broader context of interconnected colonial outposts across the oceans.

The Dutch West India Company's possessions in the colonies were connected through its imperial metropole, Amsterdam, to the Dutch East Indies. This meant that the confluence of worldviews, lifestyles, colonial imaginaries, and experiences with indigenous peoples influencing colonists in the New Netherlands exceeded the landscape of what would become New York City.[9] People arriving in New Amsterdam in the early seventeenth century quite possibly had a traveler's knowledge of other Dutch colonies through the maritime cultures of trade and colonial expansion, particularly Brazil, Batavia, and the East Indies, if not actual encounters with peoples and languages of other races and cultures.

The remarkable travel notebooks of the Dutch traveler Jan Huyghen van Linschoten, published in 1596, well before the discovery of New Netherlands' New Amsterdam, is a rich assemblage of early Dutch perceptions about indigenous peoples across Dutch, Spanish, and Portuguese colonies. Linschoten lived in India for five years, working for the Portuguese. He traveled extensively through the Portuguese East Indies, of which Goa, Cochin, and Quilon were major ports, as well as Portuguese and Dutch Africa. His detailed descriptions of the habits and customs of the people of Goa and Malabar, and the landscape they inhabited, reveal the struggle for meaning and power entailed in the colonial encounter between colonizer and native.

Linschoten took detailed notes. He traveled to Madagascar, the Azores, Guinea, Mozambique, San Thome, Ascension, Kilwa, Zanzibar, and Malindi, among other towns in Africa, as well as through Ceylon, Malacca, Borneo, and China, before returning to Amsterdam and publishing his influential journals and controversial maps of Portuguese nautical sea charts.[10] In his writings, Linschoten observes with meticulous detail indigenous cultural practices across different colonial outposts of the Lusophone and Dutch empires, describing spices, gems, minerals, fish, and plants, including medicinal varieties.[11] By the end of the sixteenth century, Linschoten's journals drew an expansive, interconnected picture of a transforming global landscape of colonial conquest and its resulting syncretic transformations.

The earliest recorded descriptions of Dutch encounters with the landscape and inhabitants of the land of the "Manhatas" after its discovery in 1609 offer yet another window into this emerging global subject of the early seventeenth century. A letter from Isaak de Rasieres, an employee of

Peter Minuit, to his senior colleague back in Amsterdam documents his surrounding environs with a cartographer's precision: "The Island of the Manhatas extends two miles in length along the Mauritse river, from the point where the Fort 'New Amsterdam' is building. . . . The small fort, New Amsterdam, . . . is situated on a point opposite to Noten Island; the channel between is a gun shot wide."[12]

Rasieres is aware of his participation in a new historic encounter. The detailed letter communicates his effort to comprehend the social world of the indigenous peoples he encounters and he offers a context that is simultaneously cosmopolitan in its references and provincial in its conjectures: "These tribes of savages have all a government. The men in general are rather tall, well proportioned in their limbs, and of an orange color, like the Brazilians."[13]

Rasieres assumes his intended readers back in Amsterdam share prior knowledge of the natives of Brazil, where the Dutch had established an earlier colony with great brutality. His comparison of the Algonquins to the Brazilians recurs in many subsequent narratives by other travelers, leading to an increasing conflation between the indigenous peoples of the New York region and the indigenous peoples of Brazil.

Rasieres's comparison is echoed in the journals of Johannes Megapolensis, for example, the first clergyman to Beverwyck, now modern Albany. Working closely with the Mohawk Indians around Albany, Megapolensis writes the earliest known documentation of the Mohawks of the New Netherlands and their social mores. According to Susan Henry, curator of the Museum of the City of New York, Megapolensis's accounts of the Mohawk peoples conflates descriptions of the Mohawks with those of indigenous peoples of Brazil. On the subject of childbirth and labor, Megapolensis observes: "The women, when they have been delivered, go about immediately afterwards, and be it ever so cold it makes no difference, they wash themselves and the young child in the river or the snow. They will not lie down (for they say that if they did they should soon die), but keep going about."[14]

Megapolensis's influential Een kort ontwerp, van de Mahakvase Indianen in Nieuw Nederlandt (*A short sketch of the Mohawk Indians in New Netherland*) (1644) became a defining text for how native peoples in the New Netherland region were perceived under the Dutch. His details of Mohawk life reappear in the work of other chroniclers of the period, such as David Pietersz de Vries and, later, Edmund Bailey O'Callaghan.[15]

Megapolensis's Christianizing mission was much broader, however. He authored an epistle for Holy Communion while working among the Mohawks in Albany. He sent the text with his wife via the West India Directors to Amsterdam, with the intention of printing and distributing his epistle in Brazil and other colonies. Apparently Megapolensis presumed that Indians in Brazil were similar to Indians in the Hudson River Valley in their need to convert to Christianity. The notion that a text created for the conversion of the Mohawks would work for other "savages," as the Dutch referred to indigenous peoples, is one possible reason for the dissemination of Megapolensis's epistle.

Curator Susan Henry's point is interesting in light of the many descriptions of indigenous peoples by early Dutch colonial travelers. The journals of John Nieuhoff, a Dutch employee of the Dutch West India Company, written between 1640 and 1649, offer one instance of elaborate documentation through travel notes of the manners, customs, and cultures of peoples across Dutch and Portuguese territories. In Nieuhoff's visual depictions of indigenous peoples in Brazil, Malabar, and Batavia, the reader sees a narrative collapsing these entirely different cultures into the notion of a homogeneous native. His detailed observation of herbs, animals, and spices, however, show careful variations.[16]

The arrival in New Amsterdam of another Dutchman and avid traveler and chronicler of the period, David Pietersz de Vries, was preceded by extensive travels in the East Indies. De Vries's ethnographic journals demonstrate how notions of borders and geographies were expanding and overlapping by the mid-seventeenth century. De Vries undertook six major voyages, the last three to New Netherlands. Following journeys to Newfoundland, the Mediterranean, and extensive travels through the East Indies, De Vries lived in New Netherlands for ten years and published his travel journals, the *Korte historiael ende journal notes*, in 1655.

In De Vries's detailed journals, the world is a small place. The East Indies serves as an influential comparative framework by which to measure the developments of New Amsterdam as a colonial entrepôt. De Vries viewed the Dutch East India Company's possessions in the East Indies as a successful colonial enterprise that shaped his critical response to the poor state of affairs in the beleaguered Dutch colony of New Amsterdam, under the effete management of the Dutch West India Company. His impatience with the inefficiency of New Amsterdam's management is set against the scenario of a more organized Dutch colonial administration in the East Indies.[17]

One striking aspect of De Vries's historical writings is the oceanic imaginary that structures his worldview. His chronicles are influenced by the violence he experienced between the English and the Dutch in the East Indies, particularly in Batavia and the Coromandel Coast. De Vries often compares the sloppy management of the young New Amsterdam colony to the more established VOC outposts he lived in across the oceans, including those in Guiana, Surinam, and Curacao: "We were surprised that the West India Company would send such fools into this country, who knew nothing, except to drink: that they could not come to be assistants in the East Indies . . . In the East Indies, no one was appointed governor, unless he had first had long service, and was found to be fit for it."[18]

This comparative framework structures his entire relationship to New Amsterdam and its uncertain future. For De Vries, the British are not to be entertained or negotiated with. They devour everything: "The English committed some excesses against us [the Dutch] in the East Indies . . . I had no good opinion of that nation . . . they thought everything belonged to them."[19] The experience he had gained with the British in the East Indies warned of what would unfold in New Amsterdam: the usurpation of Dutch territory by the British.[20]

Peter Stuyvesant, the most prominent inhabitant of New Amsterdam, honed his skills as a colonial administrator in the critical Dutch port of Curacao, whose fort was also called New Amsterdam.[21] From Batavia to Galle to Mauritius to Cape Town to Curacao, from Pernambuco to Mannahatta, a preliminary vernacular of world citizens traveling as missionaries, slaves, and indentured laborers alongside Dutch, Portuguese, Spanish, and British seafarers established a widely dispersed New World semantics. A combination of old maritime outposts of colonial motherlands—Amsterdam, Lisbon, Cartagena, and Liverpool—precipitated emerging, syncretic port encounters and wove a fragmenting reality across different maritime empires.[22]

These distinctive points of cultural collision offered threads of interconnectivity through trade routes, slave traffic, colonial administrations, and ports, as well as fort cities that resembled each other in architectural logic, while being vastly different in cultural specificity. Fort Cochin, in Cochin; Fort Aguado, in Panjim; Fort Kaapstad, in Cape Town; the Dutch forts of Batticaloa and Galle, in Sri Lanka; the elaborate forts across the Cape Coast of Ghana, with names like Fort Sao Jorge de Mina (Elmina), Fort Oranje, Fort St. Sebastian, Fort Goede Hoop, Fort Amsterdam (Cormantine); and Fort Gorée, in Senegal, were only some of these points of

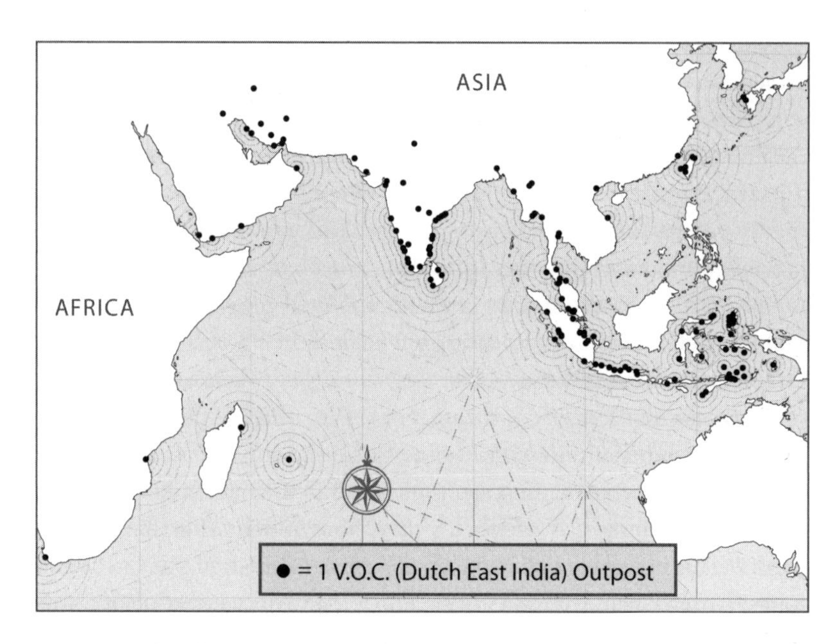

MAP 3 *Map of Dutch East India Company Combined Colonial Occupation.* Courtesy of Daniel Hetteix.

encounter. Other forts, like Barakuthi (Bangladesh); Olinde, in Pernambuco (Brazil); Thangasseri Fort, Quilon (India); Malacca (Malaysia); Batavia (Indonesia); Deshima (Japan); San Tome; Macao; Tainan (Taiwan); Mauritius; and Surinam were connecting ports in a vast network of trade and colonial occupation under the Dutch East India Company, of which New Amsterdam became one site (see map 3).

Like their colonial counterparts, the Dutch reproduced what they knew best, their beloved, reclaimed landscape of Amsterdam. They did this by giving their new colonial undertakings names like "Niew Hollandia," a name given to modern Australia by early Dutch expeditionists, and "Fort Amsterdam," a name given to many Dutch forts, including one on the Ghanaian coast, also known as Cormatine under the British, as well as Fort Amsterdam, in Curacao, Fort Amsterdam, in Surinam, and the town of New Amsterdam, in Berbice, Guiana, where the Dutch traded in slaves, sugar, cacao, and cotton. Hence, the New Netherlands in North America was not the only New Netherlands marked on early colonial maps. And "Amsterdam" was generously bestowed upon other African and Asian colonial sites under the aegis of the Dutch East India Company, as suggested by "Fort Amsterdam," in Ghana.

Johannes Vingboons's Idealized Projections

New Amsterdam's vast oceanic connection was finally brought home to Manhattan in September 2009, when the South Street Seaport Museum exhibited rare maps by the prolific Dutch cartographer Johannes Vingboons, for its exhibit New Amsterdam: The Island at the Center of the World. Johannes Vingboons was the Dutch East India Company's resident cartographer. Vingboons never left Amsterdam, but he drew detailed descriptions and idealized projections of the key ports of the Dutch East India Company's extensive possessions.

Vingboons's idealized renderings of Dutch colonial forts, which appear beside his colorful and optimistic rendering of New Amsterdam, dramatically stage the oceanic history through which New York City's past has been forged. Vingboons's 1665 maps of Ambon, Moluccas, Mauritsstad, Pernambuco, Havana, Fort San Felipe de Morro, Port Rico, Sao Tome, Elmina Castle, and Cape of Good Hope imply an interwoven history of south-to-south transoceanic connections. His vivid depictions of tropical paradises challenge the popular perception of New York's early Dutch years as a relationship largely between Europe and New Amsterdam.

Positioning New York's history against the landscape of Johannes Vingboons's maps forces an oceanic interpretation of Manhattan's past that includes the Pacific and Indian Oceans. It resituates the "island at the center of the world" against the backdrop of the economically more important ports of Ambon, Pernambuco, Batavia, São Tomé, Cape of Good Hope, and Cochin under Dutch colonial expansion. The narrative of New York's centrality is thereby dislodged—the city is revealed as a mere peripheral trading town in a provincial part of the sixteenth-century global imaginary.

The Treaty of Breda, 1667

Amid Vingboons's maps on display at the South Street Seaport Maritime Museum is a brown wax seal measuring about four inches wide, decorated with silver and red India cotton thread. It lies attached to an original document sealed in 1667, the Treaty of Breda, signed in Breda, Netherlands, to herald the end of the Second Anglo-Dutch War of 1665–67. The English had earlier seized Surinam from the Dutch, and the Treaty of Breda restored the Dutch control over Surinam, a colony the Dutch greatly valued for its sugar production. The English retained New Amsterdam, which they had also usurped from the Dutch. Skirmishes between the Dutch and the

English for Surinam and New Amsterdam suggest that New Amsterdam's fate was connected to the Dutch West India Company's better-known trading networks encompassing Asian, Latin American, and African maritime trade in nutmeg, pepper, cloves, sugar, and ivory.[23]

Elided History

The island city of Manhattan has had many identities: New Amsterdam's fort city, Dutch trading city, British colonial city, first American city, first capital city, immigrant city, port city, city of finance, city of dreams, Gotham.[24] Out of these many visualizations, the island of Manhattan rises searchingly, vulnerable in its Icarian flight, a symptom of modernity's ambivalence. Arrogant and nervous, vertical yet grounded, its dialectic of being and becoming enacts what most fascinates and terrifies us as adventurers of the urban. Its locale as a port city imbues the island with its protean quality, simultaneously entrenched and whimsical, provincial and enthralling.

What makes New York City significant as a city of world history is its superimposed connections between the Atlantic, the Pacific, the Mediterranean, the Caribbean, the North Sea, and the Indian Ocean, networks of maritime economies.[25] It remains a geopolitical nerve in the invention of modern urbanism. New York's metropolitan ways are at once passé and invite the new. Ever a microcosm of the larger experiment in urban living, the city brutally forges ahead, remembering and forgetting, between history and amnesia, an improvisation in the constant process of rewriting itself.

As New York celebrates the Amsterdam-New York cultural encounter that dates back to 1609, with extensive funding from the Dutch Government in 2009, a question raises its head: Why celebrate the "discovery" of one of the smaller ports from the VOC's vast acquisitions? New York City considers May 24, 1626, the official founding of the city of New Amsterdam. However, in 2009 the increased public discourse on the Dutch influence in the creation of modern New York has created a sense of 1609 as the psychological founding of the city of New Amsterdam. Hudson's discovery of Mannahatta in 1609 became an opportunity to revive New York's Dutch heritage, by the British presence.

At the Museum of the City of New York, in honor of the Quadricentennial, four different exhibits examine the impact of Dutch visuality, cartography, planning, commerce, and horticulturalism on New York City. The exhibits Dutch Seen, Amsterdam/New Amsterdam, Mannahatta/Manhattan, and Timescapes independently offer a wide-ranging set of connections between present-day New York and seventeenth-century Dutch colonial influences. They explore New York City across the historic spectrum of four hundred years.

At the Museum of Modern Art, an exhibition on the Dutch performance artist Aernout Mik highlights the shared culture of trading floors, stock exchanges, Wall Street crashes, and the raft of financial tsunamis that rock cities like Manhattan and Amsterdam.

At the Metropolitan Museum of Art, an exhibition of early Dutch representations of New York (including a drawing produced in 1656 by Adriaen van der Donck of the Manhattan landscape) is on display. During New York City's Summer Streets program in August 2009, a gift of one hundred Dutch bicycles in brightly colored orange were donated by the Hudson 400 organization to further the experiment in expanding alternative transportation across New York City.

On September 1, 2009, a flotilla of eleven Dutch barges in the style of Henry Hudson's era floats down the Hudson River after having sailed from Amsterdam in June 2009. Some of these barges actually have the insignia of the VOC embossed on their sails. All the ceremonial largesse is partially generated by the government of the Netherlands in its promotion of Dutch heritage in New York City and Albany.

What strikes the observer of these developments is the resounding silence surrounding the extensive cultural impact of Asia and Africa on the invention of New Amsterdam by the Dutch East India Company. One wonders: Is the Netherlands also celebrating its encounters with its more prized former possessions in Africa and Asia?

The colonial theaters in Asia and Africa were older stomping grounds for the Dutch.[26] They had honed their skills at colonial warfare and fort-building in their incursions into Asia and Africa. Their interest in Mannahatta was originally to find a way to the Orient. Hence, the reductionist cultural festivities of a monocultural Dutch influence on a largely Anglo-European New York culture elides the interconnectedness, heterogeneity, and cultural violence that the Dutch East India Company and its western offshoot, the Dutch West India Company, catalyzed across the world during the 1600s.[27] Vingboons's maps of the VOC's possessions suggest a more interdependent tale of violent exchange, hybridity, and cultural sloughing.

Maritime Mentalities

In an exhaustive study of the cultures of the Mediterranean, the historian Fernand Braudel identifies a set of social understandings and cultural behaviors he calls "mentalities." These mentalities, Braudel suggests, arise out of the conjunctions of a region connected by water.[28] Braudel argues that the Mediterranean is a region linked through the intermingling of geographically varied cultures. According to Braudel, the proximity of places and peoples interwoven through the history of water-bound travel has generated today's Mediterranean world.

Braudel's emphasis on the role of water in shaping a different kind of regional imagining introduces the part hydrology plays in the articulation of cities and regional identities. The oceanic categories of the Atlantic, the Pacific, the Indian Ocean, and the Mediterranean have respectively spawned hydrographic notions of culture that in turn shaped discursive ways of understanding social spaces. However, these disparate categories offer models of particular cultural formations: the Atlantic being the most familiar organizing metaphor for the Eastern seaboard of the United States.

The case of New York City forces a more fluid depiction of merging oceanic flows. Beginning with the outer boroughs of New York City's shoreline in Coney Island, Far Rockaway, and Sandy Hook (that incredible New Jersey isthmus that holds the entryway into New York City's harbor), and traversing into the inner waterways of the Hudson River, the East River, and the Harlem River, one is confronted with a panoply of water-defined cultures transplanted to the geography of New York City. West Indians in Queens and Flatbush, the beachfront Italian and Irish communities around Jacob Riis Park, and African immigrants in Harlem and the Bronx suggest the confluence of different oceanic cultures—the Mediterranean, the Caribbean, the Atlantic, and that of the Indian Ocean. Taking the Seven Train and the E train to Queens, one traverses the Silk Route, the Spice Route, and the Trans-Siberian pathways of Genghis Khan as one encounters Middle Eastern, Central Asian, South Asian, and Eastern European New Yorkers alongside Greek, Argentinean, Columbian, and Syrian New Yorkers. The Indian Ocean and South China Seas become visual cartographies to map the unfolding cultural pathways extending from the East River to the Atlantic Ocean on Roosevelt Avenue in Queens.[29] New York City's harbor is the unifying crucible for these vastly different communities' sense of belonging in this immigrant city. Many in the city share an immediate, if not distant, memory of journeys across oceans to arrive at the port of New York.

The oceanic paradigms of how different continental flows take place puncture the history of the invention of New York City on a deeper level. Suddenly, this city of piers, canals, slips, and waterfronts is a city in conversation with other port cities connected by the colonial history of the Dutch East and West India Companies, as well as other maritime trading posts under Portuguese, British, French, and Spanish imperialisms. The land-bound discourses of belonging to the United States and to New York City through the language of citizenship is interrupted with oceanic notions.

These powerfully imaginative constructs draw attention to contemporary flows of peoples and histories to the North American land mass that are less known here, but that occupy prominent roles in the history of New York City dwellers, such as nurses, doormen, cabbies, vendors, deli store owners, porn shop dealers, hoteliers, and motel owners. To live in New York is to be swept into the depths of a transoceanic current coursing within this maritime city.

The River That Flows Both Ways

Mahicantuck, or "the river that flows both ways," is the Lenape people's name for what eventually became known as the Hudson River.[30] The descriptive title refers to the deep structure of the Hudson River's flows—its peculiar undulation of upstream and downstream river flows.[31] The unique confluence of the East River, the Harlem River, and the Hudson River creates the unusual tides around the Hudson River. Coursing upriver toward Troy twice a day and flowing downriver twice a day, the Hudson River is the greatest hydrological experiment in North America. Commemoration ceremonies in 2009 marked the four-hundred-year anniversary of the Hudson River's usage in global trade and commerce.[32]

The naming of the river that flows both ways is a tale of the region's colonization. In a letter to one Samuel Blommaert, written in 1627, an early inhabitant of Notten Island (now Governors Island), Isaak de Rasieres writes: "On the 27th of July, Anno 1626, by the help of God, I arrived with the ship 'The Arms of Amsterdam,' before the Bay of the great Mauritse River."[33] "Mauritse River" was the name initially given the Mahicantuck by the first Dutch settlers. Named after Prince Maurice of Orange, the Mauritse was shortly thereafter called the Noort Rivier.[34]

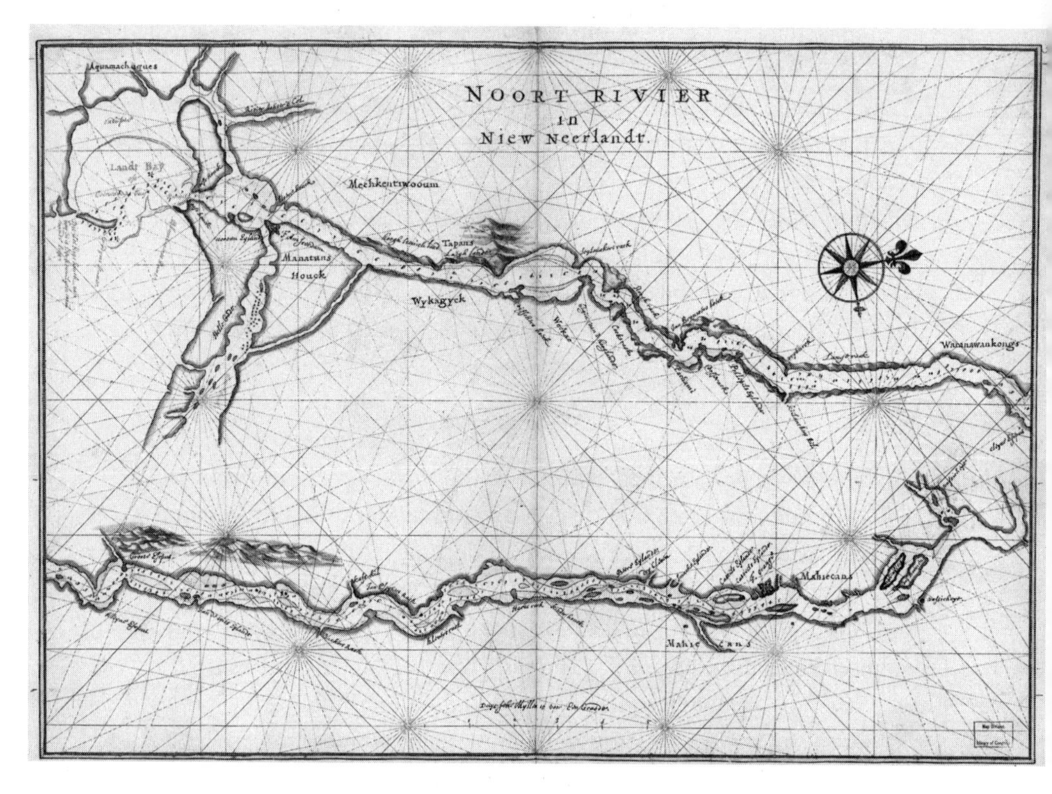

FIGURE 2.1 Joan Vingboons, *1639 Noort Rivier in Niew Neerlandt.* Courtesy of the Library of Congress.

The Minuit Chart, ascribed to Peter Minuit, depicting "Manatuns Houck," or Manhattan, in 1630, the first extant sounding chart of the Noort River suggests the early moment of reconnaissance and possibility for what would become the Hudson River. Peter Minuit's chart shows the first documented markings of the Noort Rivier's depths, done with wooden sounding lines made of a rope and weight.[35] Sounding maps were required to gauge the depths of uncharted coastal hydrography. This early map's jagged coastline etchings marking the river's depths indicate a probing expedition moving slowly along unknown shorelines, documenting the coast in the early years of New Amsterdam (see figure 2.1).

Fine lines etched across the Mauritse River's length speak to what early settlers were discovering about the river's landscape: an undulating river edge against a populated hilly island, warranting a fort at the confluence of the North and East Rivers. The serrated shoreline delineations offer a visual connection to other sounding maps deployed by colonial cartographers of the time, in particular those of Captain James Cook. Cook's sounding maps of the western coast of Niew Hollandia (Australia) show how soundings were frequently visual markers of the limits of what expeditionists actually knew.[36] Sounding maps charted shorelines without delineating what lay beyond the boundaries of the actual shore. To the extent the technology of sounding was implemented to measure depths of new colonial coastal fronts, it became a connecting metaphor for the transoceanic logic linking different regions of the world through a particular scientific practice that was distinctly ocean bound.

A River of Many Names

In A Description of New Netherland (1655), Adriaen van der Donck writes "Of the North River": "The river is the most renowned and its environs the most populous of any in New Netherland, as several settlements and the city of New Amsterdam on Manhattan Island are situated along its banks."[37] Known as the North River under the Dutch, the Mahicantuck was renamed the Hudson River by the English. This new name is inscribed for the first time in Jacques Cortelyou's map titled A Description of the Towne of Mannados: Or, New Amsterdam: As It Was in September 1661 (1664).

In Cortelyou's map, which is also known as The Duke's Plan, the Hudson River is marked against the city of New Amsterdam, even though the British had renamed the city New York by the time the map was actually drawn by English draftsmen in 1664, based on Cortelyou's survey. The juxtaposition of the name "Hudson River" and the town of New Amsterdam in The Duke's Plan captures the historic moment of rupture between Dutch colonial rule and the emergence of British hegemony in Manhattan.[38] The map asserts the superiority of British maritime power over the little Dutch colony, however, as New Amsterdam's shoreline is filled with British military vessels.

North River

The North River, as the Hudson River continues to be called by mariners and nautical charts, is a river of mythic proportions. Its narrative grandeur exceeds its physical scale. Writing in the Hudson River Almanac of September 15–22, 2009, Tom Lake, the Hudson River Estuary Program naturalist, observes: "This was the defining week in Hudson's voyage up the river 400 years ago, the week when he discovered . . . the reality that this watershed would not provide an unobstructed passage west."[39] Instead, this uniquely placed body of water would generate a world-altering set of engineering feats: the creation of a manmade canal, the Erie Canal; the execution of the largest aqueduct system in North America; and the emergence of the most prominent port of its time, the port of New York.

The *Hudson River E-Almanac*, edited by Tom Lake for the Hudson River Estuary Program, is one barometer of the Hudson River's vivacity and health. Organized as a weekly collection of observations about ecology around and on the Hudson River, the almanac registers the minute shifts of marine ecology along the Hudson River estuary. Detailed descriptions of fish, flora, and fauna shed light on changing environmental impacts on the highly populated river.

The *Hudson River E-Almanac* chronicles the biodiversity that sustains the Hudson River estuary. The river encompasses multiple ecological locations, from tidal wetlands to woods, rivulets, bays, dams, mountains, creeks, marshes, sandy beaches, and rocky shores. Its complex ecology demands its own system of referencing, the Hudson River Mile. According to the *Hudson River E-Almanac*: "The Hudson is measured north from Hudson River Mile 0 at the Battery at the southern tip of Manhattan. The George Washington Bridge is at HRM 12, the Tappan Zee 28, Bear Mountain 47, Beacon-Newburgh 62 . . . and the Federal Dam at Troy, the head of tidewater, at 153."[40] The river's measuring system and its attentive ecological watchdogs create an identifiable Hudson River habitat view. This circumscribed local ecology is modulated by the dramatic turns of tide, temperature, and weather. The resulting impact of the shifts in the Hudson River's tidal rhythms affects the river's way of life along the length of its flow.

Rights to the River

Capturing the promise of this still little-known river under Dutch hegemony, the famed mapmaker Johannes Vingboons drew an idealized rendering of the emerging port city with its serpentine river of promise, a map titled Noort rivier in Niew Neerlandt, 1639.[41]

Vingboons's romantic rendering is made to please Dutch investors in Amsterdam, who were removed from New Amsterdam's decaying state of disrepair. In Vingboons's lively watercolor, the Noort Rivier is a booming waterway. A remarkable facet of Vingboons's painting is the city's projected vibrancy and elegance as a major seaport, almost anticipating its future success.

Living by the Hudson River is to live in an interdependent ecology of water-bound concerns. The implications of residing around large, interconnected bodies of water inform imperatives from the earliest settlers. Van der Donck, a native of the Dutch city of Breda, eloquently observes: "In New Netherland many fine waterways are to be found—streams, brooks, and creeks that are navigable . . . bays, inlets, . . . as well as many watercourses, streams, and running creeks with many beautiful waterfalls . . . There are many . . . harbors, and coves . . . near . . . the North River."[42] Van der Donck's cartographic details anticipate the centrality of waterways in New Amsterdam's future. Writing of the body of water that would become Lake Ontario, he notes, "The Lake of the Iroquois . . . is as big as the Mediterranean Sea."[43]

Van der Donck's identification of the Hudson River system as a key aspect of New Netherlands in 1655 was fully articulated by the 1970s. In 1966, the deplorable pollution of the Hudson River drew together a group of fishermen and environmental activists concerned with the issue of the contaminated Hudson River watershed. They organized themselves into an advocacy body that would monitor the health of the river and become known as the Hudson Riverkeepers. Vociferous efforts by the Hudson Riverkeepers to politicize the rights to clean water and air in the Hudson River Valley and its environs dramatically altered the Hudson River's presence as a resource to be used efficiently.[44] For the first time since Van der Donck's time, the Hudson River was recognized as a vulnerable ecosystem whose habitats were under threat.

The establishment of the Hudson Riverkeepers and the Hudson River Estuary Program has led to increased public education regarding sustainable life along the Hudson. As the emergence of the online newsletter the Hudson River Almanac attests, communal engagement in tracking and observing the local ecology in the Hudson River Valley is making possible a richer history of the river's shifts and turns in flow (see map 4). The sedimentation of the Hudson River's ecology is a part of the daily life of people who live by the river. "At least a thousand ducks and geese dotted the shallow bay off Crawbuckle beach. In the midst of several hundred Canada geese, although by themselves, were three snow geese with black wingtips and snowy white bodies," Tom Lake observes in the Hudson River Almanac from Croton Bay, New York, on September 18, 2009.

The year 2009 saw the emergence of the Hudson River as a major site of spectacle. A number of planning initiatives allowed a new and sensational scale of Manhattan to emerge for the first time since the even more elaborate Hudson-Fulton New York celebrations of 1909.[45] The completion of the Hudson River Park and bike paths, along with the continued development of the Riverside Park South, opened up Manhattan's waterfront to the city. Cycling from Brooklyn across the Brooklyn Bridge to the Hudson River and then straight up the bike path to the Little Red Lighthouse, under the George Washington Bridge, near the northern west tip of Manhattan, reveals a New York never experienced before the construction of the bike paths and the greenways on the west side of Manhattan.

The city is now available to its topographically different neighborhoods as a waterfront city, with waves lapping against its rocky edges on the Upper West Side. Biking upriver any given evening, one sees large numbers of people walking, lounging, meditating, or reclining on giant boulders whose edges sink into the Hudson River. Children stroll with parents. Drumming communities gather to practice their beats at dusk. In Harlem on summer evenings at the 125th Street pier, the mood is one of leisure and picnics by the waterfront. People fish while families relax on grassy lawns. Commuters traverse the length of the island, partaking in a new culture of biking to work. This is a New York whose famed river once again offers a pleasant respite from the existentialist landscape of motorways and harried masses.

Impending Threats to the Hudson River Estuary

The cleanup of the Hudson River remains a textbook case of activist success against old and impending threats to clean water in the United States. The Clean Water Act of 1972, along with the vigilance of the Hudson Riverkeepers, the Hudson River Estuary Program, and the United States Geological Survey's Cooperative Water Program, which monitors the Hudson River Salt Front, opened up the Hudson River as a place of refuge, enjoyment, productivity, and futurity for the river's communities, both marine and human. The slow reclamation of marshlands, rivers, and bays is rewarded by a return of rare birds and insects, such as wood-eating gribbles and shipworms, according to the Hudson River Almanac.

Few New Yorkers today share memories of a bustling Hudson River filled with heavy shipping industry and maritime commerce. Tales of a New York harbor flourishing with oysters are distant history, as the oyster beds around New York had all but disappeared by 1927.[46] Corroding oyster beds became a symbol for the dying waters of the Hudson River, rank with

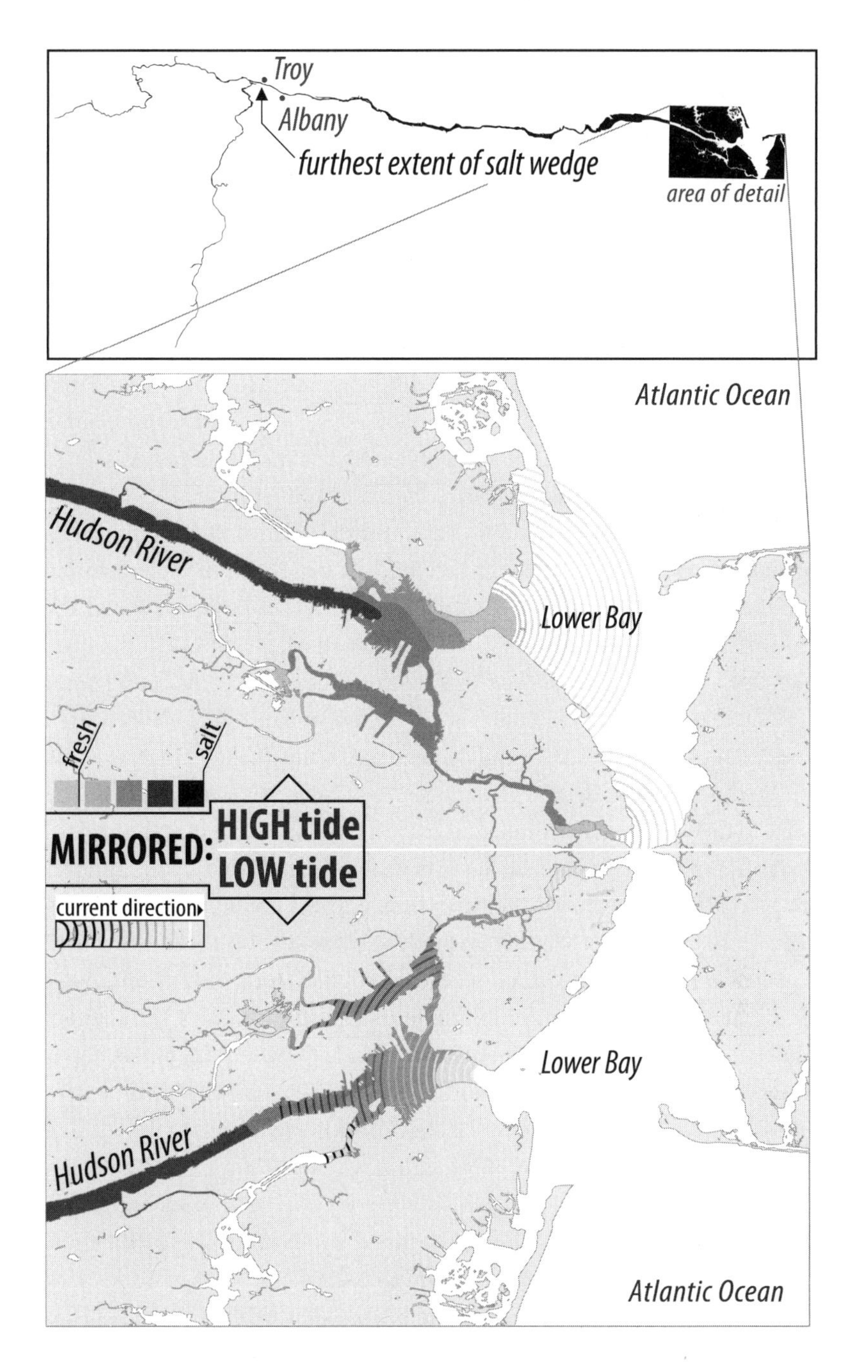

MAP 4 *Map of Salinity and Tide of the Hudson River.* Courtesy of Daniel Hetteix.

metals, solvents, oil, polychlorinated biphenyls (PCBs), asbestos, Agent Orange, DDT, and other pollutants.[47]

Once an unapproachable aspect of New York's hydrography, the Hudson River now represents an ongoing commitment to the city's future. Following decades of fights against corporate entities along the Hudson River system that were dumping toxic waste and urban refuse, the Hudson River's environments are gradually reclaiming their vitality. The oysters are returning, though are still unfit for human consumption.[48] Wind and water power are being explored. People have taken to swimming in the Hudson again, after a century of disengagement with the river. From measuring the quality of the water to the creation of a uniquely panoramic contemporary waterfront, the Hudson enters today's consciousness as an approachable presence, a river of rejuvenation and reprieve.

However, as Charles Duhigg has recently pointed out in his series of articles on "Toxic Waters" in the *New York Times*, the illusion that raw sewage is no longer released into the heavily trafficked waterways of the Hudson River and its environs is merely that, an illusion.[49] Near the Robbins Reef, in the middle of New York harbor, a large sewer discharge point for Bayonne, New Jersey, continues to contaminate the city's vicinity.[50] Furthermore, the Hudson riverbed is suffused with embedded carcinogens and toxins that have settled at the bottom of the river. The dilemma as to what to do with this environmental hazard continues to haunt the region. So far the approach has been to leave the contaminants undisturbed and let the sludge wash itself out to sea over time. But the potential for the pollutants to be disturbed and swell up into the water is always there.

Such continuing threats and the erosion of hard-won protections under the Clean Water Act of 1972, beginning in President Clinton's era, have led to the most challenging moment in the history of the struggle for clean water in America, and in New York City. Rising toxicity levels in waters across the United States, and in the New York City environs, have dispelled the idea that the Hudson River is as safe as we like to think it is. The rising new prospect of drilling for gas in six counties upstate that constitute the New York watershed is bringing to the fore the renewed threat of chemical contaminants, toxic seepage, and unimaginable danger to New York City's water supply. All of these scenarios make the stewardship of the Hudson River a critical long-term investment for the region's viability.

CHAPTER 3

the maritime sky of manhattan

I write sitting in the old Customs House, down at the tip of Manhattan, now the National Museum of the American Indian. The building, a sweeping, beaux-arts stone structure, sits squarely on the site of the footprint of the old Dutch Fort Amsterdam, built by the Dutch West India Company and later rebuilt as the British Fort St. George. The historic import of the old Customs House, as the former locale of the first settlement of the Dutch colonial city of New Amsterdam, is marked by its current public role as a national testimonial to the decimation of the American Indian. The site, framed by the green of Bowling Green Park in front of it and water all around it, is a physical reminder that the invention of the city of Manhattan is a drama of the confluence of rivers and oceans, a theatrics of water.

Manhattan was founded by the encounter between different cultures. Colliding histories of natives and travelers, of indigenous island dwellers and mainland peoples from Europe and Africa, epitomized by the first seafaring Dutch travelers to arrive on the island and their African slaves, have shaped the island's coming into modernity.[1] The city's origins lie in what has become an archetypal tale of a cultural misreading of gift exchange between a barter economy and a mercantile economy, between the non-European and the European.[2]

I imagine the first documented encounter in September 1609 between the Lenape Indians of Mannatus and Henry Hudson, an experienced British expeditionist and employee of the expansive Dutch maritime corporation, the Dutch East India Company. According to the Dutch historian Jaap Jacobs, this first encounter could have occurred around Gravesend Bay, near Coney Island.[3] Jacobs records the experience: "So we sent our Boate to sound, and found that it was a very good Harbour . . . The country is full of great and tall Oakes."[4]

On this, his third voyage, but his first to "Noort Rivier," or the North River, Hudson is nervous enough not to disembark on the island of Mannahatta. He dispatches his crew to explore the environs. Hudson's first mate, Robert Juet, writes: "The eleventh, was faire and very hot weather. At one of the clocke in the after-noone, wee weighed and went into the River, the wind at South South-West, little winde. . . . The people of the Countrey came aboard of us, making shew of loue, and gave us Tobacco and Indian Wheat and departed for that night: But we durst not trust them."[5]

In these early journal entries, Juet describes with delight what Le Corbusier would later evocatively call "the maritime sky" of Manhattan, and he notes with a mariner's precision details about the sounding of the harbor around the island. The plenitude of sea and verdant land Hudson beholds is only undercut by the expedition's suspicion of the area's natives.[6] Despite their generosity to the visitors, the Lenape are treated with great distrust. Wary of the natives, Hudson sails past the island, up the Hudson River to what is today Hudson, New York. Hudson then sails back downriver, passing the island of Mannahatta, and returns to the Netherlands.[7] Juet's journal suggests that Henry Hudson did not set foot on Mannahatta.

Sixteen years later, on May 4, 1626, the Dutch, headed by Peter Minuit, approached the hilly shores of this narrow, verdant island peopled with brown-skinned natives. Minuit met these cautious natives of the "hilly place" at today's Inwood Hill Park and initiated an exchange of gifts.[8] On May 24, 1626, he purchased the island of Manhattan from the Lenape Indians with goods to "the value of" sixty Dutch guilders. The worldview of the Lenape Indians at the time does not incorporate notions of buying or selling land in the modern sense of individual proprietorship.[9] For the Indians, this "sale" of land they didn't own is a symbolic, temporary transaction. To the Dutch, it is a real estate purchase that will evolve into the most influential artificial environment for human interaction in the world.

Drawing upon documented evidence of other land transactions made around that time, Robert Shorto, a historian focusing on the history of New York, suggests that such a ceremonial purchase would have involved an elaborate set of exchanges of various gifts, possibly food and other victuals, and feasts for the group of Indians doing business with the Dutch. Such land sales would have included the exchange of goods days before the actual day of the sale, and would continue for days, weeks, even years following such a handing over of land. Shorto points out that none of the real estate acquisitions by the Dutch were remotely straightforward. From the surviving documentation of land purchases during these early years of encounter, the Indians who "sold" the island to the Dutch had every intention of continuing to live on the land, and they did so for as long as they could.[10]

Thus begins the story of the sale of the island of Mannahatta. The transaction marks the beginning of an entirely modern experiment called American cosmopolitanism, at once a collision, a pragmatic transaction, and a performance of misreadings that together create a distinctive, evolving vernacular of living in a dense, urban environment.

This historically new improvisation enacted by people of diverse cultures and relationships to the land, people coexisting on a finite narrow land mass, is an epic tale—a tale borne of the self-preserving good will of the indigenous dweller; the frugality of the world traveler, willing or forced, if one was a slave, to keep all desire on hold; and the raw curiosity of the adventurer to the New World, seduced by the promise of the new and hemmed in by the narrow confines of the island city.[11] This founding narrative was cocreated by seafaring, canal-building people who knew how to reclaim land from the sea, as had been done by the sixteenth-century Dutch. This narrative was created by polyglot Atlantic Creole Africans from Angola and the Congo, who came via Brazil and Curacao, and who were cosmopolitan before they were free. It was created no less by island peoples blessed with a hilly home full of rivers and streams, large ponds, and pleasant bays rich in oysters and pearls. Beavers and otters lie under the water-bound imaginary of this icon of urbanism, Manhattan.[12]

My thoughts wander, and I ask: Were there East Indians from India on board these early ships as they arrived in New Netherlands, possibly after visiting more established colonies in Batavia, Cochin, Cape of Good Hope, and Pernambuco? Are they distant ancestors whose experience I in some way share? The early entries of migrants to New Netherlands describe a "servant," a "boy," and "le Orient" as nameless, faceless domestics accompanying early Dutch families.[13] Surely these pepper-and-clove-loving Dutch must have had Indonesian or East Indian labor to man their ships as well.[14] Here lies the imagined space of cosmopolitan citizenship, the space between the arrival and the invention of the city of New York.

The Limits and Exclusions of Cosmopolitan Belonging

To live in Manhattan is to live vulnerably. Knowing they live on an island engenders a distinctive urban sense among Manhattan's denizens, as opposed to those from the greater city of New York. We in Manhattan inhabit the provincial environs of a colonial trading outpost established by the Dutch West India Company, whose echoes remain in the name of Beaver Street, near Wall Street, in downtown Manhattan. We inhabit the Manhattan of the Algonquins, the Lenape, and the Munsee Indians, whose ghostly traces have been built over as lower Broadway and Gansevoort Street, Minetta Lane and Minetta Street, and the coastal locales of Canarsie and Far Rockaway. We inhabit the Manhattan of early Dutch urban planning, whose northern borders at one time were bound by today's Wall Street, with Trinity Church on the western corner and Pearl Street on its eastern boundary. That early settlement now makes up only 5 percent of Manhattan's present total land area. The historic wall that originally distinguished today's Wall Street was painstakingly built by the first enslaved Africans on Manhattan, in what one could argue was the city's first major public works project.

We also inhabit the Manhattan of older immigrants, European in extraction, hailing from colonies and metropoles. The Manhattan of the colonial Dutch and British periods shapes the public imagination of the city through the iconic representations of its gateways, bridges, canals, parks, taverns, and civic displays. The immigrant Manhattan of Coney Island's electric lights, of the Statue of Liberty and Ellis Island, of Castle Garden and the piers of maritime Manhattan reminds us of travelers to the New World disembarking nervously, excitedly, hopefully, to meet their futures.

A less familiar African Manhattan also reverberates with the maritime flows of Manhattan. This is the Manhattan of free black traders and seamen who lived on Mannahatta before it became New Amsterdam, in 1626. The first peoples of African descent on Manhattan were free men. One of them, Jan Rodrigues, a seaman with a gift for languages, was deposited on Mannahatta in 1613 by the Dutch sea captain Thijs Volckertsz Mossel. Rodrigues became an independent trader on the island, sometimes working as a translator for the Dutch and Native Americans.[15]

Other explorers of African descent also ventured into Mannahatta's environs before the formal "purchase" of the island by the Dutch, in 1626. Esteban Gómez was the earliest recorded individual of African descent to explore what is today New York. He was a polyglot and traveled widely as a seasoned member of Ferdinand Magellan's circumnavigation of the globe, in 1520. Another traveler and free agent of African descent, one Mathieu Da Costa, was famed for his knowledge of Native American languages in 1607 and was sought after as a translator by both the French and the Dutch.[16]

The history of Manhattan's black diaspora has finally been documented as part of New York City's cosmopolitan membership through a groundbreaking exhibit at the New York Historical Society in 2005, titled Slavery in New York. Historians identify the first slaves on Manhattan as Atlantic Creoles. These people of African descent bore names such as Simon Congo, Francisco Cartagena, Jan Creoli, Anthony Portuguese, and Paulo d'Angola. They draw attention to the heterogeneity of the black dispersal to the New World.[17] Their world was a complex geographical convergence of metropole and colonies. Ethnically heterogeneous and culturally diverse, these kidnapped individuals were forced to endure journeys across the Atlantic that included Dutch, French, Spanish, and Portuguese slave ports, before being shipped to ports on the mainland, of which New York was the most prominent during the seventeenth century.[18]

The area owned by freed African slaves on the island of Manhattan, under Dutch ownership in the mid-seventeenth century, was at one point referred to as "Little Africa." Black-owned territory extended north-south between Bleecker and Spring Streets and west-east between West and Lafayette Streets, before the usurpation of the territory by British forces in the seventeenth century.[19] British laws restricted inheritance and transference of ownership to relatives of freed slaves, keeping black Manhattanites slaves for another century, till the abolishment of slavery in 1840.[20] But this first group of African New Yorkers to own property in Manhattan exemplifies the founding heterogeneity that shapes the improvised political cosmopolitanism of New York City. As diasporic world citizens of myriad detours, they established lives amid different languages, cultures, histories, and belief systems.

The violent, bloody trajectory of Manhattan's cosmopolitan identity is powerfully embodied by the African burial ground north of Wall Street, extending all the way from the old boundaries of New Amsterdam, which ended at Wall Street, to the area now known as SoHo (for "south of Houston Street"), where over twenty thousand people of African descent are estimated to be buried. The continuing discovery of bodies in sites under excavation for construction, most recently in the much-disputed SoHo Trump Tower location, on the corner of Varick and Spring Streets, foregrounds the tangible history of black dispersal shaping the struggle for political citizenship in the metropolis. These material remnants are testaments to the strategies of exclusion and erasure contained within the city's vision of cosmopolitan potentiality. They delineate the limits of political inclusivity of an earlier maritime era in which the history of New York's Native Americans, the Lenape and Munsee Indians, add yet another layer of multiple exclusions.[21]

Among the newer, less newsworthy arrivals are those immigrants who rode by boat and in the backs of trucks, hid in ship holds, or ran across the border. Immigrants from the Caribbean, China, and Latin America, Jews and Christians, Muslims and Buddhists, Confucians and Taoists, Hindus and Sikhs, Cao Daists, Zoroastrians, Bahaians, Gnostics, and Atheists all jostle on the streets of Manhattan for a piece of that inchoate promise, the now withered American dream, in our time shrunk to a dystopian nightmare of sprawl and unsustainable modernity, myopia, and paranoia.

The surveillance cameras in Washington Square Park and major thoroughfares of New York City today, as well as the transient communities of the pier culture on the West Side of Manhattan around Christopher Street, gesture toward this urban history of soliciting desire and fear in downtown Manhattan. This aspect of the American myth gains ongoing life in the Manhattan of Christopher Street, between Washington and Greenwich Streets, formerly the site of Newgate Prison, the first New York State penitentiary of modern-day penology, where inmates lived and worked in from 1797 to 1828. The neighborhood's ambiance of surveillance, incarceration, and illicit pleasure continues to structure the dynamics of downtown social exchange in its many manifestations, of which the infamous and now demolished Tombs prison, near City Hall, was a physical reminder. Christopher Street's waterfront culture embodies all the confluences of utopian desire and fraught civic commitments, between public displays of forbidden seductions and historically privileged lifestyles of the big city.

The etchings of maritime Manhattan remain in the broad lip of South Street Seaport and Old Slip. The Belgian ballast stone structure of Stone Street, and Water Street on the east side of the island, where the ground continues to gradually sink into the water, are stark reminders of the older history of Manhattan. On the west side of the city, the façades of fast-disappearing buildings on Hudson Street, Greenwich Street, Washington Street, and Christopher Street still bear the signature of the city's fading waterfront industry. Most dramatically, the slurry wall of Ground Zero, with its 150-foot-wide tie-backs pegged into the bedrock of Manhattan, reminds the city of that membrane of nineteenth-century engineering that keeps the Hudson River from engulfing the giant "bathtub" most recently dominated by the substructure of the Twin Towers of the World Trade Center.

Manhattan's island structure, with its streams and rivers now built over, is most poignantly remembered as a necropolis. New York's early black communities were forced to bury their dead in undesirable areas such as marshes, swamps, and wetlands. Of historical interest in the water-bound histories of colonial Manhattan is the vast expanse of land immediately north of Wall Street, which inconveniently reminds the city of its apartheid laws, which forbade black Manhattanites from burying their dead within the city walls below Wall Street, during the Dutch and early British rule. Today's City Hall area bears the trace of New York's slave history. Its physical cartography is embossed on Manhattan's topography. Located at the busy intersection of the United States Court House, the New York County Supreme Court, the Court of International Trade, and the Thurgood Marshall Federal Courthouse, the former Collect Pond once stood, around which seventeenth-century black Manhattanites and later, Irish, Italian, and Chinese immigrants settled, to form the nineteenth-century slum of Five Points and its neighboring Chinatown, on Doyer and Pell Streets.

Poignantly, there remains the necropolitan imaginary: the Manhattan of the dead. Downtown Manhattan was the site of numerous encounters between the colonists and the Algonquins during the early days of colonial conquest, between 1609 and 1776. Twenty thousand Africans lie buried between Wall Street and Canal Street. The African Burial Ground National Monument, on the corner of Duane and Reade Streets, is a hard-fought memorial to that New York past. A Manhattan of dead indigents lies buried in former potters' fields now converted to parks such as Liberty Park, Washington Square Park, and Bryant Park. Here too lies the Manhattan of the yellow fever victims of the early eighteenth century and the victims of subsequent Draft Riots. A darker Manhattan of the grotesque lynchings of its black citizenry in Washington Square Park also haunts the city. The unscrupulous Manhattan of the Triangle Shirtwaist Factory tragedy, when 146 women were hideously burned to death, locked in their sweatshops on Waverly Place, continues to remind us of workers' rights. And there is the Manhattan of the first Jewish Cemetery in New York City, at St. James Place, where the first Portuguese-speaking Jewish immigrants to Manhattan were buried outside the city walls, though within their own designated cemetery.

Finally, there is the Manhattan that inspired Le Corbusier's invocation of the "maritime sky of Manhattan," with its relentless verticality and impatience with history. This Manhattan, with its rising skylines along its eleven avenues, is a water-bound organism in perpetual flux, transformation, adaptation, regeneration.[22] It is this resilient Manhattan that was replaced by the Manhattan of September 11, 2001, an imploding Manhattan whose ricocheting effects continue to ripple through the everyday networks of the city today, in the form of critically ill firefighters dying from lung ailments related to the rescue efforts at Ground Zero, and in the form of neural pathologies and unspoken anxieties around sonic explosions, teetering cranes, low-flying planes, and smoking high-rises.

Against this mise-en-scène of ship and sail, plane and freight, the present-day speed and density of arrivals and departures acquire a depth resonant with the inception of the island of Manhattan. People have always arrived here from somewhere else, the local inhabitants of the island having been brought, driven out, or decimated over a brief period of time during the early colonial occupation. This violent ur-text informs narratives of arrival and citizenship in New York City, the ongoing story of New York's fluid urbanism.

Movement and the City

I am riding a crowded A train, headed from Greenwich Village to Brooklyn in the rush of a midday Manhattan. As we hurtle over the gradually transforming ruins of Ground Zero, I nervously lurch along with the train and notice the sloughed-off remnants of diasporic belonging that surround me. The largely "Third World" train of black peoples from around the world, as is only possible on the A train in New York City, helps me realize that in some ways the plasma of the Indian Ocean still envelops my experience of the city, a diasporic memory whose definable trace is etched on the faces of these New Yorkers headed for the peripheries and backwaters of the city.

The boundaries of the Indian Ocean encompass the world in ways that defining cartographic lines do not. These Afro-Arab-Asian connections weave a dense set of interconnected maps and routes—the fabric of New York City. Arab grocers, South Asian cabbies, Caribbean nurses and doormen, African merchants, African American, Asian, Mexican, and Latino entrepreneurs. Afro-Arab food and Afro-Asian spices trail glimpses of the past on the streets. I inhale the scents of Arabia sold by the black Israelites in Union Square, while the Egyptian and Lebanese food vendors downtown prepare taboo food, like hotdogs, for a non-Muslim clientele. The sheer tenacity of this forgiving city's "inauthentic" citizens is impressive. People reinvent themselves and others around them in the wake of a public catastrophe, jostling for a new vision of hope.[23]

Movement in Urban Space

Cities are animate spaces.[24] Commuters, workers, vendors, laborers, tourists, migrants, and the great mass of eight million inhabitants traverse the metropolis of New York City every day. These inhabitants occupy urban space in distinctly stylized ways, consciously or unconsciously. Grasped in the gridlock of traffic and crowds, city dwellers frantically maneuver intense proximities and collisions. They activate a sphere of life called metropolitanism that is particular to the dreams and possibilities of each large metropolitan region.[25] In New York City, moving among people, things, gravities, sentiments, seductions, hailings, and distractions, the urbanite must perforce be a consummate performer and willful amnesiac. So ensconced within movement is life in the city that the elaborate and delicate art of how we move in cities, and how cities affect us, is a constantly emerging set of possibilities.[26]

From the controlled, brisk walk of the commuter across Grand Central Station to the string-leashed toddlers on the corners of streets, steadied on their way across the busy avenues, metropolitan actions are public gestures coordinated as group events while being distinctly individualized performances. Our daily acts make up the grinding underbelly of the vertical city.[27]

Propelled by the relentless push beyond the nineteenth-century idea of the metropolis, metropolitanism, as a condition of being in cities, grows increasingly intricate in its lived reality. As a growing sphere of experience, it is little understood and taken for granted as a state of being. Its intricacy is devoured by the enormity of the idea of the city and dissipated by the distraction of the urban experience.[28] It is at once an overexposed notion, a perpetually emerging sphere of being, and an ongoing arena of live practice informed by spatial movement. How people cope with movement, and how movement invents place, impacts the idea of metropolitanism.

Metropolitan movement is movement in transition. Movement catalyzed by the built environment reflects the potential of urban spaces to activate the spirit of a city. To traverse New York City is to initiate a set of spatial practices in relation to speed and mobility. How people comport themselves in particular contexts, why they occupy the particular space they do in the manner they do, demands continuous scrutiny.[29] The city is a space of negotiation as much as a place of local dialects. The tourist encounters the street vendor. The local resident brushes shoulders with the transient corporate entrepreneur. These encounters between peoples, physicalities, and spaces create microdialects of performative languages from block to block, from street to street. In the span of a stroll, individuals transition from sensibility to sensibility, as asphalt, concrete, flesh, and locomotion cumulatively generate a localness to microdialects of movement.

Movements within built environments are collaborative productions of muscular activities and civic understandings. Improvisations of movement within cities generate the distinctness associated with the sensibility of cities. A language of the street unfolds between the pace of the pedestrian and the inhabitant of the in-between spaces of the street.[30] These local dialects of bodily gestures articulate a sphere of exchange that sets the feel of the neighborhood. Hence, walking on Fifth Avenue feels quite different from walking on the Upper West Side, or in Fort Greene. The woman strolling along Washington Square Park encounters a rapid transitioning of pace as she hits Bleecker Street, only to shift to a more leisurely gear as she veers off toward the calm of Leroy Street or Grove Street. Each type of city dweller needs pedestrian life to produce stylized actions particular to spatially specific contexts: vendors on Canal Street, idle cabbies, the cigarette-breakers in the Financial District, art dealers outside the Metropolitan Museum of Art, demonstrators at Union Square, horses sipping water in Central Park.

Metropolitan spaces are spaces of conjuncture. Chinese bodyworkers on the streets of Manhattan collide with pedestrians, while Pakistani, Sikh, Sudanese, and Egyptian cabbies syncretize their aesthetic of driving in Third World arenas as they transport the rider along. They redefine New York's streets, making them theirs. In turn, New York's streets reflect the extraordinary diversity of its cabbie culture, so a cab ride becomes an immediate intimate space of metropolitan encounter. African American bus drivers, alongside Russian, Haitian, Nigerian, Iranian, and Bulgarian driving styles, merge a mosaic of road logics honed by the intensity of driving in New York. This semantics of danger and proximity encapsulates the lives held on hold in the interim space of a city.

Metropolitan spaces are spaces of unexpected movement, where habit and density merge with rehearsed perambulatory gestures. They collide with other movement regimens. The movement of bodies takes on the shape of structures around them as large groups of people meander through concrete, glass, and steel. Bodies accustomed to suburban spaces gingerly tread the dense sidewalks of metropolitan life, often dressed inappropriately in impractical suburban footwear, or flamboyantly conspicuous shoes that collapse under the strain of New York's unforgiving macadam. These encounters allow for the flexibility of the unexpected in a ritualized culture of walking, viewing, seeing, and ignoring.

Spatial Dwellings

For powerfully evocative and globally dispersed cities such as New York City, this compressed flow of movement raises the underlying question: How do people dwell in such cities of distraction? "What is the nature of dwelling in our precious age?" asks Heidegger, writing of a time when the shape of cities and the density of bodies within cities forced a perpetual concern of the modern era, the concern for housing. "On all sides we hear talk about the housing shortage . . . the real plight of dwelling does not lie merely in a lack of houses. The real dwelling plight lies in this . . . that mortals . . . must ever learn to dwell."[31] This not inconsiderable challenge of urban modernity manifests itself through the escalating tension between movement, bodies, and the built environment.

In New York City, the architectonics of habitation are incommensurate with the heterogeneity of bodies.[32] Multitudes of every nationality straddle the asphalt daily, as the barometer of real estate reconfigures flows of inhabitants across the city's interiorities. People hug the crevices of the city as would bipedal vermin. They perambulate to variegated rhythms of money, labor, desire, and pressure. How these multitudes "dwell" in their temporality, in their ephemeral permanence as urban dwellers, is the imaginative realm of "dwelling" at its most unfettered. This dwelling is the space where movement and habitation merge to create the urban experience in motion. It is the infinitely pleasurable and inchoate space of urban life that distinguishes walking cities from other kinds of cities. Such a dwelling is the space of intimacy between people and the built environment that allows for an animated interchange of sensuous experience. It becomes an incomplete, unfolding space of urban possibility. Dwelling, movement, and events are simultaneous vectors across which urban spatiality is constituted.

Event Urbanism

Drawing on perceptual regimes other than the visual to describe the particular nature of buildings, the architectural theorist Bernard Tschumi describes the building as an event—a process where spatial practices, performative actions, and physical encounters are realized. For Tschumi, movement is fundamental to the built environment, as much as the circulatory system of blood and veins is central to social interaction.[33] The event, that interactive space of action and exchange, is the symbiosis of body and building, the animate space of concrete and flesh entwined through the imaginative leap of design and sentient experience. Events are transient acts of dwelling through which movement is created. They invoke the relationship between action, space, and the human form.

Cities are conglomerates of events, for Tschumi. This notion of event extends the performative dimension of action as a "happening" to a set of possibilities broader than the dimension of the human form and more malleable than the conventional use of performance space. To live in cities is to occupy the spaces of multiple events: physical, somatic, and built. Extending Tschumi's conception of event to the spaces of the city, not merely its buildings, one sees that events are encounters between the built, the living, and the material. They are casual, everyday, and banal as much as they can be exceptional, spectacular, and soul-satisfying. Cities generate events, are composed of events, and eventually become events.

Conflictual Urbanisms

Metropolitan spaces are spaces of repetitive gestures. The modulated body movements of the urban dweller counter the space-consuming mannerisms of out-of-towners. The neighborhood street dweller stages his gigs for the itinerant visitor, and hustling local businesses enact their deals for gullible visitors eager for a memento to mark their foray into a city they do not really like or care to be in for more than a few days.

Metropolitan spaces are often stages for the unconscious rejection of dense living by Americans fearful of cities. The city is the anomalous place of encounter where the plenitude of suburban ease and space has to be sacrificed for a few days of intense confirmation as to why cities are undesirable places. The mentality of "I like to shop in Manhattan but I need my SUV, my suburban home, and my backyard" spills into the spaces of the city, as visitors inhabit the city in passing. The street is filled with people who pass through but do not live in the moment of their encounter with the street that disconcerts them.

In the twenty-first century, as more people gravitate toward cities for economic reasons, metropolitan spaces become compressed stages of nonmetropolitan desires in transition. These spaces become chasms of disenchantment. The spaces of Manhattan are replete with these nonmetropolitan desires, as the gentrification of the last two decades has seen suburban needs and commodities etched on the compact urban landscape.[34] Increasingly, suburban lifestyles are infusing urban dwellings with extravagant comfort zones of discrete luxury. The tiny, dingy railroad apartments that once housed ten or more members of a family are being transformed into open, airy, spacious apartments through the fusing of two or three apartments. Low- and middle-income dwelling places have been priced out of New York City, in particular around Manhattan, to make room for those who can pay the exorbitant rents to slum it in New York City. Manhattan in the first decade of the twenty-first century emerges as an unimaginably wealthy borough, dramatically displacing the very populations that make the city flow in its innocuous and unimpressive dailiness.

The systematic pricing-out of the city's lower- and middle-class populations is dramatically shifting the demographic and cultural potentiality of Manhattan toward an economically homogeneous population of affluent urban dwellers. As gas prices surge, exurbias are becoming the new city, and the inner city is transforming into the new suburb. The Lower East Side projects are now middle- and upper-income condos. Harlem has transitioned into a desirable area, outpricing its own residents in the process.

Neighborhoods like Red Hook, Atlantic Yards, West Chelsea, Chinatown, the Meatpacking District, the Bowery, and Greenpoint are some of the areas where this marked shift toward a new visual coastline is emerging in New York City, accentuating the city's unspoken economic disparities. The terms "affordable housing" and "low-income housing" are catch phrases used in the development stages of many of these proposed waterfront properties. However, their visualization and design do not invite the diversity of economic stratas to which they initially promised to reach out. Brooklyn Bridge Park is an instance of such an effort to bypass Brooklyn Heights zoning and historic preservation in order to establish luxury condos in the park. The park's envisioned mixed-use utopia of spiral pools and landscaped pier parks, combining different income levels, is a picture yet to unfold, as the upscale One Brooklyn Bridge Condos looks for renters in a down-market economy.

The Coney Island Redevelopment Plan is perhaps the most watched proposal being tested. Floated designs suggest mixed-use and mixed-income demographics occupying the area. However, the implied high-end residential buildings that are planned for some of the waterfront sites will create certain forms of visual exclusivity through the aid of guards, design elements, broad plazas, and complicated entranceways, thereby limiting access to the Coney Island beach over time. These are only some of the problems anticipated by interested community watchdogs. It will be critical to New York's future imagining to realize a fully democratic and publicly accessible Coney Island waterfront free from the threat of a shadow of towers looming over this historic beachfront of New York City.

Greening Metropolitan Movement

Alongside the contentious privatization of iconic public spaces such as Coney Island a new kind of metropolitan movement is being activated in New York. It is a kinetic panoply of physical and technological mechanisms whose vocabulary includes a wide range of alternative transportation networks geared toward creating a more livable, pedestrian, and accessible city. This impetus toward a livable metropolitan streetscape has been particularly tangible under the stewardship of Janette Sadik-Khan, the Department of Transportation's commissioner under Mayor Michael Bloomberg.

Discussions about greening metropolitan mobility under the Bloomberg administration has included the creation of over four hundred miles of bike lanes, more pedestrian access to waterfronts, bridges, and the shoreline of the city. The pilot projects of closed off streets along Union Square, Thirty-Fourth Street, Broadway, and Times Square is permitting new kinds of rhythms to emerge in the city. A decidedly new pace of slowed-down traffic is physically and aesthetically restructuring how people inhabit streetscapes.

Traffic-calming techniques involving landscaping and visual cues to reduce the speed of traffic, along with the generation of attractive, small-scale plazas along hectic streets such as Broadway, are innovations that have introduced New York to a new kind of urban leisure lounging in the midst of intensely busy intersections such as Harold Square and Times Square. The technique of traffic calming, unfamiliar to New York, is being experimented with amid considerable public outcry from bus drivers, cabbies, business owners, and commuters at large. Despite resistance from drivers, these new techniques to reduce noise, accidents, and improve community life along busy junctions such as Fourteenth Street, Thirty-Fourth Street, and Forty-Second Street are cultivating a discernible public shift toward a new concept of metropolitan movement.

A further push in the direction of the greening metropolitan movement is the public talk of expanding commuter ferry services to reconnect New York's five boroughs to each other, as well as to New Jersey. Santiago Calatrava's proposal to create aerial gondolas on Governors Island for the transportation of people between the island, Brooklyn, and Manhattan has opened up discussions about other ways of creating transportation fluidity. Other discussions addressing the need to connect downtown Manhattan to JFK Airport, as well as to Newark Airport, through direct rail links, as a means of creating more fluid travel flows, suggests that the larger picture of metropolitan mobility is beginning to emerge in a new framework of speed and efficiency. These last two discussions are largely visionary at this stage, but they present the conflictual potential of fluid urbanism still shaping New York's metropolitan imaginary. They are ideas that open up movement on a metropolitan scale between boroughs, cities, and regions. Metropolitan movement in New York is currently a vibrant and contentious hotbed of innovations and pilot projects being tested on the public, with many new ideas still on the drawing board to be implemented toward the goal of achieving a more ecological metropolitan landscape.

CHAPTER 4

thinking metropolitanism

"We must think in metropolitan terms, not city and suburban terms," writes Thomas Bender in his book *The Unfinished City*.[1] The notion of thinking on a metropolitan scale in the twenty-first century is both an abstraction and an aesthetic practice. One sphere of metropolitan thinking is the idea of metropolitan experience itself. Today's metropolitan sensibility is compounded by the multiply distended boundaries of expanding urban centers.[2] Still, the need to immerse oneself in and capture this elusive and alluring condition continues to hold urban dwellers in thrall. For Bender, the current historical moment calls for a "new metropolitanism" that grapples with the unmanageable scale of the contemporary city, with its decentralized circuits of flow, ambiguous borders, and myriad urban clusters that constitute the metropolis. Such a metropolitanism departs from the experience of the modernist city with a definable core and coherent boundaries.

Bender's call to think in metropolitan terms is a challenge increasingly framing local discourses in New York City. A growing sense of New York's own metropolitan identity has shaped an ecological awareness of place and locality in the city's sense of itself since the 1970s.[3] Imagining a metropolitan scale blown to its blurry edges, exposing its vastness, its ungraspable contours, has opened up different questions of urban self-invention.

Living the metropolitan scale entails thinking in metropolitan terms.

The physicality of daily life translates scale into a sensation coursing through one's physiology, a sentiment expressed on a sidewalk. For the urban dweller, magnifying the scale of New York to its outer edges has immediate tactile and sensorial implications as she traverses the tunnels and labyrinths of metropolitan density. Yet, the language of metropolitanism eludes daily parlance. As a discourse, metropolitanism permeates city life without translating its intimacies into a vernacular of dailiness. This evasive condition of the new metropolitanism is constantly reinventing New York City as it convulses, erases, rebuilds, and realigns its metropolitan imaginaries in new directions. What continues to fascinate inhabitants of the city is the metropolitan experience.

Montage Metropolitanism

Fractured metropolis. Delirious metropolis. Vulnerable metropolis. Maritime metropolis. Numerous are the perceived sensations of the island city of Manhattan. What remains constant is a notion of the early twentieth-century city, a towering metropolis of cyclopean canyons, at once visionary and blinded by hubris. It is Le Corbusier's "savage" city.[4] It invokes Fritz Lang's futuristic vertical city of *Metropolis* fame. This idea of the metropolis continues to inform subsequent inventions of urban modernity, from the notion of sprawl and the idea of the megalopolis of the 1950s, to the emerging formation of the edge cities of the 1980s, to the ambivalent new urban communities of the twenty-first century.[5]

To live in this city is to perpetually face the fleeting reality of urban life through its relentless leaps into the future. Everything changes in the city. This is the nostalgic observation of every New Yorker. Yet, New York is by no means unique in its relentless embrace of the urban landscape. In 1930, New York was the largest city in the world, with two hundred skyscrapers and seven million people. Today, at least fifteen world cities exceed New York's population. Most of them are in Asia, headed by Mumbai. Hong Kong had the highest density of skyscrapers in the world in 2010, with the Tokyo-Yokohama metropolitan region being the densest area in the world.[6] These gigantic urban agglomerations have the earned distinction of being called megacities. Scale is no longer a reason for New York's lure.

What distinguishes New York City from many other fast-paced cities in the world is its claim to distinction as an island city planned on a modernist utopian grid laid out in 1807, a city whose experimentations with a walking metropolitan lifestyle have inspired people to imagine themselves in the fantastical scapes of urban space, whether real or fictional, since the early nineteenth century.[7]

Manhattan's density deters any effort at segregation. Its irrepressible flows from the streets resist stagnation. The city's heartbeat pulses new forms of becoming in a landscape of perpetual change. Despite its youth in comparison to many of the great European and Asian cities like Paris, Istanbul, and Beijing, New York's provenance as the quintessential modern city of the future, offering the world new ways of imagining urban space, remains. It is an aura hard-earned through its tenacious public history as a city predicated on dense vertical construction, with spectacular vistas created by its visionary planner, Robert Moses, and the passionate struggles of its neighborhoods to insist upon maintaining a human scale to allow for the intimate pleasures of a walking city, struggles most powerfully embodied by the figure of Jane Jacobs.[8]

Admiring this ruthless ability to forget footprints of former buildings, Le Corbusier observed in the 1930s that the impressive scale and design of the disappearing Manhattan avenues built on the north-south axis were modernist works of art that forged a new visual understanding of the cityscape.[9] For Le Corbusier, historical forgetting was Manhattan's quintessential characteristic and its brutal lure. Comparing Manhattan to the Paris of the 1930s, Le Corbusier remarks that the very arrogant ability of Manhattan to relentlessly move beyond sentimentality and historic preservation is also its impetus toward its realization as the eternally changing city of the future.

Historical forgetting is indeed part of New York City's becoming, as a walk down that most iconic of streets, the Bowery, attests. The famed Bowery of flophouses and derelict buildings has transformed itself unrecognizably into a suburban-style housing complex with upscale stores to attend to its now wealthy million-dollar-condo denizens. What is happening to the Bowery is perhaps the most dramatic testament to a New York urbanism premised on historical forgetting, despite the urban activist Jane Jacobs's powerful advocacy in the 1960s against the erasure of neighborhoods in the interest of characterless suburban structures.[10]

Jacobs's efforts during the 1960s to save the West Village and SoHo from being razed in the face of a proposed Lower Manhattan Expressway (LoMex) forced a more people-centered approach in urban planning in New York City, despite cynicism on the part of developers and real estate corporations. However, the dramatic sprucing-up of the Bowery with lusterless buildings of single-occupancy mega-businesses, such as Whole Foods, and high-end residential developments accentuates the uneven impact of Jacobs's thinking on urban planning in New York. In the case of the Bowery, Jacobs's advocacy of historic preservation was eschewed in the transformation of the historic area into a mixed-use neighborhood of commercial, residential, and entertainment activities. One can witness the workings of the city of historic forgetting noted by Le Corbusier nearly a century ago. However, the reinvented Bowery has also considerably reenergized the area in new ways, drawing in droves prosperous youth, tourists, and consumers of food, culture, and entertainment. It is a more upscale and homogenized Bowery, but also a more vibrant and lively area today.

Despite the brutal hand of developers, and in tension with Le Corbusier's reluctant admiration of a city that forever seeks the future on a finite land mass, the residents of the city, along with the preservationists and urban activists of New York, desire a New York of the future that is more respectful of the sentiments and feelings of its neighborhoods and citizens. Contrary to the aura of indifference, anonymity, and isolation projected upon the city by outsiders, the media, and tourists' passing impressions, the people of the city of New York remain greatly concerned about how their city's growth remains in touch with their own daily lives.[11]

For many New Yorkers, the city's metropolitan identity has been shaped by people with democratic ideals who are invested in public life in shared public spaces.[12] This New York is a place that embraces the ideal of a tolerant mutual coexistence embedded in local identities and public memories—which frequently collide with each other.[13] It is a place that is always at the crossroads of deep immersion in its current state and futuristic speculation.

In this New York of dimly lit community meeting rooms, street protests, public hearings, local civic organizing, random exchanges at the bus stop, interfacing public spaces, public parks, and the steps of City Hall, citizens from disparate walks of life engage in discussions about how the city can evolve, while keeping the interests of diverse communities in mind. This spirited New York is the metropolitan neural workings of the city—a mechanism that is heterogeneous, democratic, and publicly invested in its identity as a world city. Unglamorous, resilient, and indomitable, this New York is a theatrical performance of civic life of a magnitude never before imagined.

American cities are interpretations, or rejections, of the dense verticality of Manhattan. Verticality is no longer the determining idea of what the city is, as Herbert Gans observed in his case study on the Levittownization of American life.[14] In Gans's study of the first suburban enclave, New Jersey's Levittown, he notes that suburban communities tend toward interiorized concerns. The private sphere becomes the focus of suburban dwellers sequestered by their single-family dwellings, cordoned off by lawns, driveways, white picket fences, two-car garages, and empty sidewalks (if there are sidewalks at all). The absence of a shared public life in a heterogeneous array of contexts leads to a less publicly engaged community life; in the suburbs the focus of living is centered on the hearth rather than the piazza, park, or street.

American experiments in urbanism tend toward what the urban geographer Edward Soja calls "spread" logic. Soja points out that most Americans live in urban contexts, but only a portion live in cities.[15] This observation delineates the distinction between living in a city and being a commuter to a city while living in a greater metropolitan area. It implies an American modernity organized around suburban, car-defined logic that privileges the three-car family over public transportation. This scenario temporarily collapsed in 2009, in the face of the rising costs of fuel and the implosion of the housing market and subprime loan mortgages.[16]

The morphing of American urbanization into what Soja terms "exopolis"—the various permutations of "edge cities," "metroplex," "outer cities," and "technoburbs"—is fracturing notions of what the city is. A proliferation of metropolitan forms abound, giving rise to what Soja calls "improbable cities where centrality is virtually ubiquitous and the solid familiarity of what we knew as urban melts into air."[17] What such a dramatic realignment of cities and resources such as oil and energy bodes for American urbanization is still unfolding. However, one scenario of a return to the vertical city is emerging.[18] More families with children are moving to downtown Manhattan, transforming single-occupancy apartments into family dwellings. Downtown Manhattan was a destination for single-occupancy dwellers during the postwar era. Current reclamations of inner-city centers and downtowns amid proclamations of the end of cities offer futures still to be imagined.

The exploding reality of Asian megacities is in tension with American urban modernity. According to the United Nations World Urbanization Prospects 2009 revision, over twenty cities exceed the population of New York's eight million inhabitants.[19] Tokyo, Delhi, São Paulo, Mumbai, Mexico City, Shanghai, Calcutta, Dhaka, Karachi, and Buenos Aires have populations of over thirteen million.

Soja observes that contemporary urbanism is being radically reinvented by the furious growth of these massive metropolitan regions.[20] Gigantic urban formations in Asia are directly impacting contemporary understandings of what life in big cities entails. In turn, these gargantuan city cultures are taking their cues from, and being shaped by, the dialogue between historic pasts and the future of urban planning across regional and national borders. The resulting debates between localization and metropolitan urbanisms continue to replay tensions between urban citizens and city governance, between the built environment and neighborhood needs.

Playing an insidious role in this international dialogue is the growing prominence of the now-global dream of American suburban modernity. America's standalone single-family dwelling, with its media room, family room, and playroom, is a floating promise of Asian modernity, despite its impracticality for dense city living. Sprawl is posed as a haven from dense urbanism. The desire for single-family dwellings still remains the ultimate middle-class dream in India, despite more efficient vertical housing strategies that privilege historic preservation, human-scale interaction, and mixed-use development. In such an expanding scenario of urbanization, the particular directions and endeavors of New York City remain of international interest, both as an exception in the American model of urbanization, due to its sheer scale and history of verticality, and because of its iconic place in the history of futuristic cities.

According to the historians Gerald Benjamin and Richard P. Nathan, New York was the nation's first great metropolitan experiment.[21] New York's identity as a metropolis transformed dramatically alongside the extraordinary expansion of other American cities along the Eastern seaboard.[22] The interconnected urban nexus of Boston, Washington, New York, and Baltimore, introduced in the 1950s, became an emerging urban ecology tenuously described by the term "megalopolis."[23] By the late twentieth century, the idea of the metropolis mutated into multiple interpretive schematics best epitomized by the splintered schema of Los Angeles's multiple urban cores and the dispersed schema of Atlanta, ever expanding outward.[24]

The "megalopolis" is an unwieldy concept. It emphasizes the conjunction of networked cities in regional systems, creating a larger set of interlocking urban frameworks than the singular notion of a self-enclosed city. The idea of the megalopolis poses difficult questions for visualizing how urbanism might look on the ground, between these interconnected corridors of human flows. The macrostructure of the megalopolis diminishes the human-scale workings of local ecologies that coexist within such a megalithic construct.[25]

Baltimore, Boston, Philadelphia, and Washington, D.C., are distinctly different urban and physical environments, from each other and from New York City, while they also share many commonalities on issues of gentrification, diversification, transportation networks, the revitalization of neighborhoods, the historic preservation of buildings and industrial sites, and the impact of water ecologies.[26] Their proximity to each other does not dilute their distinctiveness. Regional proximities sharpen conversations on what the city is, and what it could become. Such dense urban conjunctures on the Eastern seaboard activate a critical staging ground for the future of human potential. The flows of people between these cities, and the interlinking conversations on energy, sustainability, conservation, and maximizing efficiency have forced a reassessment of what life in cities and between cities might mean for the urban dweller.

In the United States, the flattening frame of the 1950s notion of the megalopolis was accompanied by the outward expansion of segregated social needs through the cultural embrace of suburbanization, a trend that also accelerated in the early 1950s.[27] The systematic dismantling of railroads across the country, from the mid-twentieth century onward, initiated a move away from a cultural investment in public transit networks.[28] It heralded the emergence of the Fordist model of car-culture efficiency in American lifestyles, epitomized by the suburban home.

This ideal, privatized dream reality, enabled by the activation of communication technologies, is juxtaposed with intensified urban encounters, vertical constructions, public transportation networks, and built environments captured by the experience called metropolitanism. More so than ever before, the need to understand how people inhabit cities, and how cities shape human encounters in the face of macrostructures of urban interconnectedness and alienation, has foregrounded the elusive idea of metropolitanism as a way of life within large urban clusters.

Thinking Metropolitanism

Modern metropolitanism is a montage of manifestoes. From Baron Haussman to Robert Moses, from the New York activist Jane Jacobs to the current New York mayor, Michael Bloomberg, modern metropolitanism's commentators have been many and contentious. What arises from the different approaches to the challenges and demands of understanding the explosion of urban densities is the shared assumption that metropolitanism is a shifting practice of everyday life connecting dense-city dwellers around the world.

Metropolitanism has become the singular informal public discourse affecting millions of migrants and local citizens in every major urban context. It is a network of imagined, concrete, legal, and physical forces that creates the gigantic reality of a metropolitan region. While metropolitanism suggests the condition of being in a singular city, the lived reality of life in today's cities is that the cultures of many cities inform life in any particular city.[29] Cities have always been agglomerations of multiple urban vernaculars.

Metropolitanism invokes big-city life with its worldly and international flair. The prefix "metro" maintains its nineteenth-century sense of a polis or city being connected by the expanding network of the metro, or train system. The word implies speed, anonymity, challenge, indifference, verticality, heterogeneity, and change. Metropolitanism is a construct of urban planning, architectural vision, transportation networks, and the regional linking of urban centers to each other, hence generating a sense of the metropolis that is expansively about the city itself. It is the set of physical, cartographic, somatic, and technological practices that comprehensively generate a culture of interconnectedness for urban centers constructed through and around large cities.

The term "metropolitanism" conveys the condition whereby cities extended through transportation networks become entities greater than their individual parts. In time, they bear a persona of habitation that is greater than their individual identities. Distinguishing today's metropolitanism from that of other eras are the increasing speed, heterogeneity, multiplicity, and palimpsestic mapping of global cultural vernaculars onto the corporeality of world cities. The high-speed rail systems at work across Chinese cities, in particular, demonstrate this intensification of density and speed on a scale unmatchable in any American city.[30] If cities have never been knowable, if their mappings have always been partial and incomplete, then today's metropolitanism is ever more intricate and discombobulating.

How an undocumented Nepalese pedicurist working on Bleecker Street for a Korean nail salon negotiates metropolitanism in her stilettos is distinctly different from how a postmodern French dancer forages for a fragile living in New York's artistic world. The pedicurist, without legal papers to work in the United States, travels from Courtelyou Street, in the outer periphery of Queens, to Bleecker Street, in Manhattan, and rejuvenates tired urban feet for a living. The dancer teaches French and works in experimental productions at Merce Cunningham's studio to fulfill a dancer's dream.

A Kenyan, Muslim al Qaeda operative being held in the federal prison in Chinatown for his role in the first World Trade bombings is incarcerated amid great secrecy before his trial in the federal courts in downtown Manhattan. Nearby, on Forsythe Street, a Chinese delivery man with a former life in a small town in Guang Dong Province occupies the space of the Manhattan street with a boredom and sense of chore that is different from the aesthetically coded lifestyle of a New York City bicycle messenger who pedals his way through dense traffic with the subcultural attitude of messenger culture. The bag, the accessoried couture, the bike, the posture, the speed, the style of cycling amid cramped vehicles are the trappings of a distinct subculture and distinguish bike messengers from other cyclists on New York's streets.

Along Hudson Street, a woman who walks dogs professionally, with ten dogs of varying sizes in tow, maneuvers around snow piles and garbage cans, while on Sixth Avenue a festively decorated Taco Van serves irresistible deep-fried Mexican street food à la Nueva York. On West Fourth Street, the baker Monsieur Claude, of Patisserie Claude, who commutes from Queens, bravely persists with his delectable assortment of pastries as one of the last merchants from the good old days of the 1970s, while Mr. Ali, the Pakistani American deli store owner on the corner of Horatio and Eighth Avenue, who lives in New Jersey, struggles to hang on to his business amid skyrocketing real estate prices.

The different patterns of engagement of these very different individuals are nuanced by their immersion in the city's rhythms of speed and dailiness. Their very disparate lives suggest that the metropolitan experience has a local ecology particular to its landscape. City living incorporates metropolitanism as an intimate practice. Metropolitanism is a local engagement, as much as it is a set of macroscale urban experiences. It is simultaneously an organic and dislocating phenomenon, made specific by the geography, locality, and fictive life of the city itself. As the habits and customs of peoples from other cities of the world converge into the compact area called New York City, the practices of metropolitanism explode into myriad possibilities. Having seen everything, but shocked by suburban insensitivity, jaded, but delighted by the unexpected antics of a toddler by the roadside, the urban dweller experiences sensations of metropolitanism that are refreshingly uneven.

Metropolitanism is an exhilarating and contradictory condition. Its semantics are cloaked in entrepreneurial risk-taking, in bold choices, in endurance and creativity. Its vernacular also includes surveillance, suspicion, doubt, paranoia, and fear. The macroscale tendencies of metropolitanism, such as large-scale commuter networks and complex relationships between working and living spaces, erode systems of cultural coherence even as they bolster practices of metropolitan life.[31] Despite its disassociating mechanisms, metropolitanism nurtures new dreams of belonging, of assimilation, and of innovation, dreams shaped by a will to imagine a different kind of future. Its culture of neighborhood politics and civic engagement continues to counterpoint the relentless privatization of the public in everyday American urban life.

Car Culture versus Human-Scale Cities

The humbling image of humans perambulating amid the scale of Manhattan's skyline powerfully dramatizes the quintessential problematic of living in New York: its scales of movement. One first encounters the towering landscape of an incredible manmade island as one hurtles in a cab toward the Triborough Bridge, now renamed the RFK Bridge. Kinetic movement veers the traveling body between concrete structures, speeding vehicles, suspended bridges, and finally ejects one into the city itself, calmly waiting at the end of the approach. Then one's feet touch the ground.

The most fundamental tension of life in New York lies in this encounter between walking and the street, walking and the vertiginous heights of the built landscape, walking and the speed of the moving objects on the streets, roads, and highways surrounding the human subject. This fundamental relationship is most powerfully captured in the towering and much revered, hated, and, these days, increasingly admired, figure of Robert Moses.

The singular, sweeping impact of Robert Moses on the island of Manhattan and its environs boggles the urbanist. Swimming today in one of the city's many massive public pools, with their public-works façades, or watching children scamper around any of the city's numerous parks, now slowly expanding, one is reminded that these instrumental civic spaces for rehabilitating the masses were initially conceived, designed, and executed by Moses in a concerted effort to build more public amenities for an urban population he privately disliked, as per the accounts of Robert Caro's *The Power Broker*. Moses's fervent desire to clean New York City's unwashed masses through public bathworks transformed the grimy urban environs of a spiraling city and led to a new public commitment to democratic leisure.

In his poignant account of the destruction of the Bronx in the wake of Robert Moses's mechanical vision, Marshall Berman, an urban historian, paints a moving picture of mid-twentieth-century metropolitanism. It is a brave and hopeful portrait, haunted by the specter of a progress predicated on forgetting. In this picture, the future of the metropolis is built over the bodies of a variety of thriving neighborhoods in the Bronx of the 1950s. This hard-edged metropolitanism of Robert Moses is alive and well today in surprising places, such as the embattled West Village, which escaped the hand of Moses, though not his legacy. The recent acceleration of luxury high rises, such as buildings designed by Richard Meier, Frank Gehry, Jean Nouvel, and Rem Koolhaas, and less persuasive architectural invasions, such as the Gansevoort Hotel and the Standard Hotel, over the High Line in the West Village, announces the shift in the visual phrasing of the New York skyline along the west side of Lower Manhattan.

The technological spectacle staged for the car, as envisioned by Moses from the RFK Bridge, the Verrazano-Narrows Bridge, and FDR Drive, has finally given way to an increasingly pedestrian-determined street ambience—a pace structured by the human gait and bicycle. Searching aggressively for an alternative way of traveling within the city's arteries, New York's residents are ardently fighting for a change in transportation perceptions and options. The current push from Transportation Alternatives and the Department of Transportation to encourage residents of the city to participate in visualizing ways of shaping road habits is encouraging. Slowly but surely, a slight shift in the mentality of the city toward an acceptance of a walking metropolitanism is under way. A dialogue about finding alternative ways of inhabiting the metropolis is embracing the city's attenuated scale and form.

Walking Metropolitanism

Urban sprawl has reached its zenith, and the energy crisis is forcing a rethinking of how we inhabit urban space. City dwellers are faced with the challenges of how to cope with our ever shrinking resources in an unsustainable, expanding culture of energy consumption. The 2008–9 global economic downturn foregrounded shrinking ecologies in an atrophying international market of consumption. The West Village, in Spring 2009, felt like a ghost town around Hudson Street and Bleecker Street, with shuttered stores and the growing presence of the homeless on benches in parks or along the waterfront, in doorways, subways, construction sites, and libraries.

In such a critical period, New York City offers an evolving register of one kind of American metropolitan experience that is both a historical and dynamic modern counterpoint to dominant American views on the future of dense living. New York's idea of metropolitan life is invested in ideas of public space and public access that are grounded in a public transport system that prioritizes a walking metropolitanism. In Jane Jacobs's neighborhood, where many eyes once surveyed the street, there are enough vacant commercial properties that one walks home late on a workday evening with a sense of ill ease. Yet walking is key to keeping the streets alive.

At a time when few Americans walk to work, school, or for leisure, the privileging of a walking urbanism on the scale that unfolds in New York City every day, involving people at rush hour, is an extraordinary spectacle that never ceases to impress. It is a performance of active metropolitanism that demonstrates what Jane Jacobs eloquently stated in her writings: that cities are for people, and should be reclaimed by people for their daily pedestrian use.

The first sensation to grip a visitor to New York after the compulsory fear brought on by the hair-raising cab drive from Kennedy Airport or La Guardia Airport through the forbidding urban jungle of Queens, or through the disconnecting industrial-urban vistas of the New Jersey Turnpike, is the impact of one's feet on the ground, when one is caught in the middle of human and motorized flow. From then on, the city alters the dweller, visitor, and commuter in intimate ways that begin with a language of the feet. To live in New York in that sense is to literally think on one's feet. It is all about feet, feet, feet. As Dr. Seuss ponderously asks, "How many different feet do you meet?"

This quintessential logic of the feet became by the 1950s the single most challenging argument against motorized vehicles and the romance of the expanding highways of America. Its most vocal and charismatic contender came in the form of a mother of three who walked her children to school every day and walked home to prepare dinner. Jane Jacobs's searing critique of America's abandonment of its interest in feet, its own human biopower as a productive commodity of city life, threw into relief the motorized fantasies of Moses's urban future.

Moses, who never drove himself, envisioned a city to be viewed from a traveling car. His vision was propelled by his passenger-seat perspective. Yet, amid the impressive, cinematic, real-life spectacles Moses hewed out of iron, steel, stone, and concrete emerges the painful reminder that movement still matters, in ways that the human eye alone cannot sustain. The city requires an embodiment grounded in human feet.

Quality-of-Life Metropolitanism

Precipitating a public discourse on metropolitan life was a series of measures implemented by Mayor Rudolph Giuliani under the dubious phrase "quality of life." Articulated as a series of zero-tolerance measures to clean the city of its decadence and depravity, the quality-of-life statutes aggravated the colliding spheres between public space and private lives, between civic regulations and individual lifestyles. These diverse and invasive articulations into what is permissible in an urban space, and what is allowable within urban lifestyles, exploded into myriad encounters between city governance and urban dwellers. Suddenly, street vendors, jaywalkers, gay clubs, cabarets, bars, street buskers, demonstrators, and bicyclists were all violators of metropolitan expectations. The very inhabitants of the metropolis and their ways of life were brought into fraught confrontations with coercive forms of disciplining.

New York's metropolitan lifestyles conflicted with New Yorkers' identities of difference. The fine line between the city as possibility and the city as crucible of surveillance blurred, triggering new unease among New York's residents. The fear of going to a gay club, of moving one's hips in a bar, of crossing the street away from the zebra markings, contaminated the pleasurable unpredictability of living in a world city.

The irony behind the quality of life injunctions was the resulting reduction of the city's quality of life. Mayor Giuliani's administration curtailed public participation in policy concerns. Blockades to City Hall on Park Row and Broadway, initiated in August 1998, along with rules restricting the press and public from the center of city politics, confined the public culture of free speech. A tangible, deepening anger, nervousness, and anxiety about social policing arose, eroding the freethinking individualism of New York's public life.

A call to reopen access to City Hall as the historical marketplace of free speech resounded across the city by late 1998. City residents demanded that the expanded roles of the police in civic life be diminished and that the 114 community gardens to be auctioned off in May 1998 be saved. The city desperately needed to overhaul its daily operations with energy-efficient technologies, and citywide calls to address the poor management of waste export and public transportation were rising. The escalation of child death from asthma had skyrocketed in New York City since the 1980s. Low-income housing was in catastrophic decline. The city's public parks were in awful disrepair. Ridership on the city's subways increased in 1997, while train services were scaled back. Giuliani's administration proposed cuts to programs for children and seniors, putting large numbers of the city's vulnerable populations at risk. Such were the actual implications of Mayor Giuliani's quality of life initiatives. It left the city's residents asking, What sort of quality of life? For whom?[32]

Bloombergian Metropolitanism

In New York, the urban citizen is deeply embroiled in the politics of recognition. This metamorphosis from merely being a person who lives in New York to an individual who participates in the life of the city of New York as a New Yorker, transpired through a series of mega-events following the World Trade Center bombings.

After 9/11 the question of metropolitanism came to the fore in a manner never before imagined in the history of American cities. Mayor Bloomberg's call to New Yorkers, on September 11, 2007, to move beyond the task of mourning and look forward invoked this new construct of the New Yorker as citizen of a city that has shared a catastrophic event. This rhetorical hailing of New Yorkers unified the identity of the city in previously undefined ways. Bloomberg's signal to move toward an engagement with New York's future was a controversial gesture aimed at changing the public rhetoric of a wounded New York metropolitanism. It was an invitation to embrace a new metropolitan mentality—to move beyond the sentiment of salvage.

This forward-moving momentum of a New York metropolitanism was articulated in the shadow of a war that was cynically launched in the name of New York City's tragedy. The city of New York has historically been protective of its own immigrants and been vociferous in its commitment to a metropolitanism that is perhaps the most impressively heterogeneous and inclusive model of metropolitan diversity in the world, in terms of sheer scale.[33] When anti-immigration legislation forced local authorities to report undocumented workers, New York City chose to not disclose the legal standing of its inhabitants working in restaurants, bodegas, and other businesses.[34]

The post-9/11 call to be suspicious of anyone who looks suspicious was posted on the city's subways, train stations, public thoroughfares, and transit points and underlined the tensions inherent within the commitment to difference. Emblazoned on subway cars, on buses, in subway terminals, and on subway maps, the disconcerting "IF YOU SEE SOMETHING, SAY SOMETHING. BE SUSPICIOUS OF ANYTHING UNATTENDED" has cultivated a culture of alarm and nervousness. These cautionary signs have in turn anesthetized inhabitants of New York against the vagaries of fear and suspicion. Under Bloomberg, however, the ambience of daily fear receded as the most objectionable of Rudolf Giuliani's quality-of-life measures were abandoned. A return to the *feel* of a less policed New York took its place. In reality, however, a more surveilled New York took shape following 9/11.

Bloomberg's metropolitanism is a fraught metropolitanism. The city has been in recovery from catastrophe. Manhattan's downtown underwent extreme cosmetic surgery. The resulting transformations of New York's façades were arguably experiences designed for affluent boroughs. Manhattan saw a dramatic new possibility of movement emerge under Bloomberg, as its waterfront transformed from abandoned piers into a circumference of greenway, enabling pedestrian circumlocution around the island for the first time. The completion of the Hudson River Park, as well as the high-concept elevated park called the High Line, grant Bloomberg credit as the mayor who greened New York most dramatically since the creation of Central Park, in 1857.[35]

New York also saw terrible losses under Bloomberg in public amenities, such as mixed-income housing and public health investments. The closing of St. Vincent's Hospital, in 2010, a venerable and much-needed institution for Manhattan's downtown residents, is a stark reminder of the setbacks New York is reeling from during Bloomberg's tenure. Instead of strengthening access to health and education, listening to the public's need for more hospitals, and taking a more neighborhood-centered approach to eminent domain, the administration has erred on the side of corporate interests and developers. The ruthless hand of top-down planning has been imposed on neighborhoods, despite the active efforts of areas such as the Brooklyn Atlantic Yards and Morningside Heights, in the Upper West Side of Manhattan, to resist the forced demarcation of eminent domain.

Under Bloomberg, mixed-income housing became more scarce, public amenities for the middle class, such as rent stabilization, lost ground, and housing estates such as Stuyvesant Town, a mixed-income development, went market rate. Hard-won protections for low- and middle-income communities eroded, leaving large swaths of New York's working populations vulnerable to the avarice of real estate markets.

Accompanying these material onslaughts on the working sector, concerted efforts were made to reclaim greenspace around New York City's boroughs, with efforts to green neighborhoods in desperate need of parks falling short, as in large parts of the Bronx, Queens, and Brooklyn. Bloomberg's metropolitanism, which linked the city through cross-borough initiatives, such as the creation of park space and bike lanes, came at a strategic time in the city's imagination. Covered in debris and ash, New York was eager to breathe again, to look for the green and the blue of the city's past. Reclaiming its foliage and waterfront became emblematic of a shift forward, begun by the Green Thumb movements in the East Village of the 1970s.

Reclaiming the waterfront allowed the city of Manhattan to become more livable under Bloomberg. Bike lanes were expanded and green spaces developed across the city, in an effort to remedy the dearth of vegetation around New York. These citywide efforts have initiated the slower work of extending these practices across all boroughs, to create access to green space for every child in New York.

For many in New York, however, Bloomberg will best be remembered as the mayor who proved that money counts. He aggressively demonstrated that one could dictate the fate of a metropolis by overturning term limits through sheer financial power.

Biking Metropolitanism

As the global economic crisis sinks into the fabric of New York City's infrastructure, and with the slashing of New York City's budget in 2009, affecting library services, traffic, law enforcement, parks, child welfare, transportation, education, housing, hunger, and the environment, the city has registered a marked increase in the use of the bicycle as a mode of transportation.[36] Cyclists have demanded a rethinking of city streets. The impact of Critical Mass, Time's Up, Transportation Alternatives, Bike New York, and a number of other groups concerned with alternative modes of transportation within urban contexts has slowly shifted the emphasis in New York City toward human movement. Parents wheeling children in Pedi cabs, inventive cycle rickshaws, tandem cyclists, unicyclists, bicyclists, electric bikes, pedestrian traffic, inline skaters, Segways, baby strollers, and runners have reclaimed Manhattan's waterfront as a place for alternative transportation. Designated bike lanes, both active and under development, along with street-calming techniques that interrupt the flow of traffic in busy thoroughfares, have opened up the city to new kinds of encounters, performances, and events in the middle of frenetic traffic.

Illusions of Green Metropolitanism

PlanNYC2030, Mayor Bloomberg's goal to plant a million trees and scale back the use of plastic bags in New York City, has been a high-profile push for the city to reimagine its environs. Transforming defunct industrial areas into green spaces, along with the reclamation of brownfields and superfund sites such as the Gowanus Canal, is one direction this metropolitan place-making has taken. However, this sartorial effort has been more cosmetic than structurally transformative.

Under Mayor Bloomberg's administration, practices of waste management have been environmentally destructive. Recycling and waste reduction was abandoned as a waste-management policy, and waste export to destinations outside the city or the state was embraced with expansionist efficacy. The unsustainable practice of expanding landfill use in upstate New York to absorb the city's solid waste was a serious setback to the Solid Waste Management Act of 1988, which identified the use of landfill as a harmful waste management option for communities and the environment. Even as more people in the city compost waste and create microgreen spaces in tight dwellings, more garbage is being generated by the city of New York. The rising interest in putting roof gardens on formerly empty rooftops, closing streets to traffic in heavily trafficked areas, such as Times Square and Herald Square, and the use of hybrid buses and cabs has initiated a discussion on how to reduce the city' carbon emissions. However, the city's biggest environmental challenge, managing its waste production, remains in a state of impending crisis. The Bloomberg administration has not adequately addressed the pleas of concerned citizens around the state of New York to curb New York's unsustainable production of garbage, which is exported to small, poor communities with little lobbying power to protect themselves against aggressive waste-management companies. The urgent role garbage plays in the greening of neighborhoods has not been seriously addressed in New York City to date.

New York has very poor planning for both residential and commercial waste. Only a small portion of New York's waste is recycled. Residential waste is sent to landfills in Newark, Ohio, South Carolina, Virginia, and Pennsylvania, and from there to more distant destinations outside the purview of environmental laws. The city's commercial and industrial waste is sent to poor, rural communities in upstate New York, whose environmental safety is greatly endangered by the influx of large amounts of solid waste. In the face of New York City's massive waste production, the region is facing an insurmountable waste-management emergency with far-reaching implications for the city's livability and land-use requirements.

Consequently, while on the surface the city is popularizing the idea of a more responsible relationship to the environment across demographics, demonstrated in thousands of public school children painting decals of flowers to decorate cabs in New York, and in Summer Streets, a car-free experiment that began in 2008 to introduce New Yorkers to the idea of a more human-scale movement defining mobility in the city without being threatened by cars, in reality, the city is in the midst of an environmental catastrophe.

This impending emergency is accompanied by a growing consciousness about sustainability and the future of dense living in the minds of many New Yorkers, as people cram into subways filled to capacity and many subway routes are terminated in the era of New York's financial crises. Longer waits and intervals between trains, along with fewer bus routes in a city of nearly ten million people, on a workday, provide fuel for rethinking livability in cities. This new awareness of a shrinking urban ecology of scaled-back transportation possibilities and increased public transport dependency is forcing a new degree of metropolitan intimacy.

Hyperlocality as Metropolitan Sensibility

The dramatic erosion of New York's newspaper culture, along with the fates of newspapers around the United States, has led to the creation of hyperlocality as a way of experiencing the city. Dedicated, site-specific local-information networks such as EveryBlock, Placeblogger, and Patch generate new relationships between people and their immediate habitats. People are informed about local unfoldings in greater detail as these new social networking pathways offer increasingly localized information. The city is shrunk to blocks and neighborhoods, fragmented into experiential segments of life, such as restaurants, crime, real estate, and streets.

The segmenting of neighborhoods into smaller audiences through new media interventions appears to be both an expression of what is taking place within neighborhoods and a catalyst for further segmentation of social life. Hyperlocal advertising is creating new local identities within streets and blocks. People are increasingly specialized in what they are looking for, as they narrow their search by geography and by topic. This mediatized production of local identities is networked through larger intersections of transportation, interborough traffic, and pedestrian encounters, to produce a form of metropolitan sensibility that shifts from block to block, and street to street.

Nuanced localization, in tandem with a flattening of metropolitan routines, makes the experience of metropolitan life a highly calibrated set of civic experiences. City dwellers don't need to engage with the city at large anymore by reading *Time Out* or the *New York Times* for their information on the city. They can just search their hyperlocal blogs for information posted by people who live in the neighborhood, as opposed to journalists and reporters who document information in more objective ways. Blurring journalism, reportage, and anecdotal postings, these new information channels have opened up unimagined avenues of detailed engagement with locality, much of it uncorroborated and frequently subjective.

For many urban dwellers, hyperlocality has evacuated a sense of the larger outline of the city as an overarching idea within urban life. People are increasingly locally absorbed in their practice of city-making. Through these new media interfaces, metropolitanism, as an experience, has splintered into highly provincial experiences of placeness, including online neighborhood communities. Notions of the looming, unmanageable city have transformed into layers of manageability, from the macro to the local and even smaller, breaking down into blocks. These new sorts of interfaces between inhabitants and the scale of their lifestyles have exploded new kinds of engagement with urban place-making that are unfolding every day.

Fractal Imaginaries

In his influential book *Postmetropolis*, the critical geographer Edward Soja presents a scenario of fractal urban formations. Soja persuasively argues that urban life in cities like New York and Los Angeles is well beyond any graspable notion of a coherent metropolitan experience. The metropolis is in crisis, and its structures of organization in perpetual states of expansion and erosion.

Soja presents a splintered and decentered world of compartmentalized urban experiences for the urban dweller, without moorings in any particular geography outside the personal. It is a profoundly disembodied metropolitan imaginary, propped up by sensations of the hyperreal and fed an intensified diet of simulated urban feelings. Soja's fractal cities are free of notions of context and locality, making ideals of neighborhood, belonging, and engagement extremely fragmented social formations at best.

Soja's theorization of a "postmetropolis" is daunting. It shrinks the human capacity to imagine alternative life worlds amid tremendous adversity and distraction. The myriad avenues, where the theater of public engagement collides with the drama of public spaces in New York City, attest to a deep desire on the part of urban dwellers to forge temporary encounters of intimacy. Through performance, protests, parades, walks in the park, or exchanges at the corner deli, streams of coherence emerge despite the chaotic distortions of the postmetropolitan condition. Despite the contemporary urban disposition toward an "exopolis," a condition Soja identifies as a "city turned inside-out," the contrary shift of "the city turned outside-in" is also under way.[37] It is not just the urbanization of suburbia that is taking place, but the globalization of inner cities. These multipronged pathways of human exchange create new kinds of urban experiences that extend beyond the category of the "metropolitan" and demand other kinds of approaches to understanding how cities like New York are fracturing and recentering in this yet-different phase of metropolitan reimagining.

What is remarkable about New York is not its hyperreality or its multitudinous subworlds. Like Los Angeles, New York has many centers, despite the centrality of Manhattan to the city's imagining. New York's particularity is its small-town mentality, its sentiment of the street and the block, a feeling structured by the logic of a walking city that throws the distortions of the exopolis into perspective. Like the moment of emerging above ground from the subway station, the moment of reentry into the streets of

New York locates the urban dweller in an immediacy of the urban experience that is distinctly local and familiar, regional and dispersed. This structure of feeling threads through the alienating effects of a fractal city. One is simultaneously within what Soja identifies as the "unbound metropolis" as well as grounded in an intimacy with the street that is bound to a phenomenological engagement with the present.[38] It is a slippery thread of connectedness that allows New Yorkers to maneuver through life with a semblance of coherence.

Metro Scales, Manageable Lives

Communication technologies that are allowing new kinds of intimacies to emerge, concretize, and transform urban spaces are proliferating despite the dystopian unfolding of urban modernities, places of increasingly discontinuous assemblages of dense living arrangements, as they are mapped by Edward Soja. New York's increasing populations are also inventing innovative social connectivities. The city is in the throes of a previously inconceivable communication transformation that is reshaping how people relate to their neighborhoods.

As New York's waterfront transforms its land-use objectives, it creates an unexplored new city along the fringes of the city's boundaries. Merging its preoccupations with water management, real estate, and historic preservation in creative and futuristic ways appears to be one of New York's visionary hopes. Increasingly, the city is witnessing planning proposals and speculative exhibitions on what can be imagined along New York's waterfront in the foreseeable future. The projected rise of the oceans by at least two feet in the next century will have serious implications for New York's downtown. Furthermore, the predictions of increasing hurricane-strength weather, along with changing weather patterns in the region, are influencing discussions on how to plan a future of increased density that also prepares for escalating sea temperatures.

Scenarios of potential floods threatening New York's historic downtown necessitate a different approach to thinking about locality and space. The city is vulnerable to climate change, and its plans for metropolitan expansion need to include long-term flood prevention. While New York is no longer a maritime city in the historic sense of the term, New York's role as a model port city preparing for a future unimagined fifty years ago endows it with a new importance that is different from its earlier prominence as a trading city with a great harbor. The city's numerous histories, as a fort city, a commercial city, and, at this point in time, a city of gardens and spectacular river vistas, factor into the conversation about what the city will become. New York's experiments in waterfront redesign and greenscaping remain interesting for their scale and successful deployment of adaptive reuse. It is a city gradually adapting to its ecological vulnerability as a daily fact.

PART II

cosmopolitan frugality

In his essay "Idea for a Universal History with a Cosmopolitan Purpose," Kant draws on the Greek Stoics and Pauline Christianity to lay out founding legal arguments for the right to the city, the right to hospitality.[1] Kant argues for a cosmopolitan citizenship that is defined by certain cities as a condition borne out of the right to refuge, and the limits of hospitality.

Picking up from Kant and the early history of refuge cities, Jacques Derrida lays out the predicament of cosmopolitan citizenship, as it entails multiple conditionalities from state to state, and city to city. Derrida points out that the ideal of cosmopolitan belonging is a potentiality set against the limits of historic articulations of cities of refuge. Some cities only allowed visitation rights, while others claimed "open" doors, within limits of economic viability.[2]

The following section, "Cosmopolitan Frugality," offers readings on the impact of political cosmopolitanism within urban life. Each chapter demonstrates principles of material practice through which public recognitions of cosmopolitan belonging are activated within the city. Together, these chapters explore the interface between traveling urban philosophies transported through migrancy and the quotidian techniques of the self that cultivate diverse possibilities for cultural and political citizenship.

The theme of frugality emerges in this section as a particularly galvanizing global principle influencing different strands of cultural engagement. The following section exposes competing global geographies of uneven development impacting New York's surface. These visual markers of a distinctively political cosmopolitanism transform the city's public spaces in imperceptible ways. The quest for an enduring and equitable notion of cosmopolitan belonging is contained within these widely divergent technologies of frugality.

CHAPTER 5

nomadic urbanism and frugality

On a hot September day in 2007, Barack Obama visited New York City's Washington Square Park to give his inaugural address for the Democratic ticket. The excitement and hope was tangible. People were festive, despite the skirmishes uptown between the Iranian head of state Mahmoud Ahmadinejad, President George Bush—whose presence had been largely ignored—and other dignitaries from the United Nations, who were meeting to discuss global warming, once again without the presence of the United States at the World Economic Forum on Climate Change. The day allowed a very New York moment regarding security detail. The presence of George W. Bush and Mahmoud Ahmadinejad uptown, and Barack Obama at Washington Square Park, encapsulated the global within the periphery of a small island. Dense proximities between multiple political imaginations were highlighted, staged in the international arena of the United Nations buildings and Washington Square Park, a historic destination for political demonstrations.

The Washington Square Arch, designed by Stanford White, with its bas reliefs of George Washington on either side, startlingly framed the complex history of migration, displacement, and citizenship that resulted in Barack Obama's being born an American—and his status now as a symbolic New Yorker. Juxtaposing the dream of a president that reflected the demographic of New York City against a public park known for hosting activist and antiwar factions was a thrilling event for the city of New York, still bedraggled by the politics of aggression and the antiwar rallies at the United Nations in 2004.

Obama fit seamlessly into the persuasions of New York City, with his demographically heterogeneous constituencies of all races and sexual orientations. The crowd downtown around Washington Square Park was young, internationally diverse, and pronouncedly populated by people of color, of all demographics, alongside the predominantly white Greenwich Village residents of today. Particularly striking was the marked presence of peoples of African descent gathered around the area known in the mid-eighteenth century as "Little Africa," to support the first black presidential candidate nominated by a major party.[1]

During the 2007 election campaign, the rights of the front-running candidates to represent New York City and, by extension, their claims to cosmopolitan citizenship were playing out in differing registers of intensity. In tension with Rudolph Giuliani's undisputed status as the mayor who rallied a wounded New York City from the scene of catastrophe and unprecedented chaos on 9/11, and Hillary Clinton's phenomenal transformation from the wife of a disgraced president to a senator from New York State, and then to a front-running candidate for the presidential nomination of the Democratic Party, with all the pedigree that a life in the White House and the aura New York gave her, Barack Obama presented a political cosmopolitanism that was indicative of the future of urban belonging. While not a native New Yorker, Obama reflected the nomadic urbanism that quintessentially captures the spirit of American cosmopolitan identity. At once African American and African, simultaneously the descendant of slave owners and postcolonial African history, Obama embodied cosmopolitan citizenship as an international legacy of encounter and reinvention. His childhood in Hawaii, ancestral connections to Kenya, kinship affiliations in Indonesia, school years in New York City, and residence on Chicago's South Side, mirrored the dramatic trajectories of urban migrations and metropolitan identities that shape world citizenship today. Obama reminded Americans that he, like the great metropolis of New York, was a citizen of the world, rather than of a region of the world. Obama effectively laid semantic claim to a universal cosmopolitanism, manifested by the city of New York itself.[2]

Metro Mentality

Time distends and the senses mutate the crucible of the metro mentality. In this city, locals, immigrants, migrants, and tourists forge provisional resting places, temporary solidarities, and fleeting protests to stage a performance of modern democracy, despite the ongoing challenges that threaten its script.

Little Tokyo, little Italy, little India, little Korea, little Vietnam, little Myanmar, little Africa, little Tehran, little Guyana, little Colombia, little Philippines, Manhattan's Chinatown, and Flushing's Chinatown all coexist within the city. A composite of local knowledge and urban vernaculars from other cities inhabit the city. The A train traverses the African American, Dominican, and Caribbean diasporas from Harlem, Washington Heights, and Flatbush, traveling to East New York and Far Rockaway. The E and Seven trains take one on a simulacrum of the ancient Silk Route displaced onto Astoria, Flushing, Forest Hills, Jamaica, Jackson Heights, Rego Park, Howard Beach, and Far Rockaway, in Queens. Greek, Chinese, and Korean New Yorkers sit cheek-by-jowl with Egyptian, Iranian, Uzbeki, Kazakhstani, Bukharan, Lebanese, Baluchi, Syrian, and Pakistani New Yorkers, as the subway hurtles through imaginary routes from Colombia to Syria, from Lhasa, Katmandu, Dhaka, and Inner Mongolia to Vladivostok. It is a route that reminds the rider of the visionary mendicant G. I. Gurdjieff's travels through Central Asia.

New York's subways are a microcosm of its larger transcontinental air and sea routes. Its waning port-city culture, with older routes in the Indian Ocean, Pacific, Caribbean, Puerto Rico, Latin America, and Africa has compounded the demographics of its earlier European diasporas.

Made up of regional urban systems alongside oceanic imaginaries, daily life in New York is a fractured provinciality whose vulnerability as an icon of modernity has been unmasked by the implosion of the Twin Towers in downtown Manhattan. The city's dailiness foregrounds the reality that most residents do not live in their place of birth, and consequently their worldliness is an assemblage of experiences. On the subway trains alone, life-worlds that are distinct in their cultural and bodily expressions intermingle. Actual routes of travel perpetually fragment the nomadic city, transform its residents, and in turn are morphed by its changing demographics. Neighborhoods absorb imagined landscapes from other regions and nations. Kabul, Istanbul, Dublin, and Rangoon are superimposed onto the interiorities of the city, alongside Zanzibar, Addis Ababa, Cairo, New Orleans, and Fez. Meanwhile, the port-city structure is unsentimentally dismantled to present a glass and steel New York, relentlessly pursuing architectural euphoria around its site of trauma downtown. Postcolonial New York fuses with a new imperial and expansionist American century.

As New York rapidly hurtles into the unprecedented building phase of the present, the historical sense longtime New Yorkers have of their right to the city is giving way to a culture of rapid-fire initiation and provisional arrangements. Established ethnic enclaves have dramatically shifted from Italian to Chinese in NoLiTa (north of Little Italy), Manhattan, for example, and from Greek to Arab in Astoria, Queens. Once ethnically entrenched areas now contain first-generation South Asians, or Koreans, or Afghanis, or Colombians. Nomadic movement, a sense of having moved many times before arriving in New York City, best captures the urban sensorium of New York today.

Nomadism is a fragmentary experience, a distinctive and blurry condition of being in the world. Historically diverse and contemporaneously generic, its cultural roots and permutations are many. Often unconscious or unplanned, the phenomenon of nomadic urbanism encapsulates the spillover from a variety of situations, including civil wars, insurgencies, suburban sprawl, and economic downsizing. Its forms range from the local, the sedentary, and the regional, to the transnational. These states of nomadic urban self-invention have degrees of invisibility, as they mutate and conform within the boundaries of the possible.

Transnational urban nomadism is an uneasy and ambiguous condition of existence, experienced by a considerable portion of the world's population. As a way of life, transnational nomadism exceeds the notions of nationalism and citizenship, but is profoundly shaped by the intersection of these two sets of constructs. It is in this space of the inarticulate—between the need to be legal in a state and the desire to create a life in a new city—that the palimpsest life of the refugee, immigrant, or sojourner gets played out. Optimistic, distracted, caught between many places, belonging nowhere, the transnational nomad is at once cosmopolitan and curiously local.

This space of the urban nomad is a place of wretchedness as much as it is a place of liberation. It is a place of unknowing and a space of projected utopias. Urban nomadism is an anonymous life hemmed in by laws, fears, uncertainties, and debts accrued in former lifetimes: a journey into the peripheries of modernity's excesses. It is a place of little rest and less ecstasy, a place of deferred dreams and desires placed on hold. It is a place of hasty leavings and incomplete farewells. This nomadism is a web of limits, where imagination, resources, hope, and the ability to endure outweigh the legality of citizenship within the nation-state.

Migrancy and Nomadism

Nomadic migrancy is an international phenomenon.[3] What distinguishes contemporary migrations and struggles for immigrant citizenship from those of earlier eras is the distinctly new sense of historicity that marked the passport-carrying subject traveling from one nation-state to another during the twentieth century. The social ramifications for urban nomads have been far-reaching, as large groups of people move across local borders and, eventually, oceans. The move from village to city, from urban peripheries to urban centers, has created new kinds of feelings, and has generated new somatic encounters with emerging experiences of modernity.[4]

Modern nomadism is a performance of self-invention. Nomadism's global landscape is forcing new arrangements of community, as digital and communication technologies open unimagined spaces of habitation. Migrancy's scripts are being rewritten and reinscribed as populations shift around the globe. Its maps are many.

There are the rhizomatic journeys of the singular nomad whose individualism allows for the distinctness of the journey traversed.[5] There are the journeys of the first postindependence travelers and migrants who never return to their place of birth, whose reasons for leaving their country of citizenship are economic, material, and phantasmatic.

There are the migrations of settlers who are searching for a better life. The settler colonies of Australia, the United States, Canada, and South Africa are societies forged out of this sort of migration.

Political and economic migrancy, in particular, force new and often conflicted relationships to homeland and return. People moving en masse across political borders transport existing political imaginaries, while also forging new social possibilities in their new place of belonging.

According to the cultural theorist Teshome Gabriel, the roots of modern nomadism in Africa lie in the traditions of its wandering Saharan and sub-Saharan tribes.[6] This idea offers a mythical and metaphorical context for modern African histories of dislocation and fragmentation. Displacements, such as illegal boat migrations, compelled by economic imperatives, are a forced leaving of a very contemporary nature. Gabriel's theory, however, links such contemporary migrations to an older practice of migration as a way of life in some African narratives. For many African boat people who take their chances on the high seas of the Atlantic and across the Strait of Gibraltar, hoping to reach Spain or Italy, migration is a deadly gamble, rife with threats of death, incarceration, and deportation.[7] These new illegal migrants have no accompanying narratives to help them cope with the waves of anti-immigration sentiment that have opened up in the European Union, compounding the challenges of dealing with illegal migrants in urban milieus across Europe.

Nomadism as American Lifestyle

Contemporary urbanity is bolstered by people who are placeless, as they travel from one job to another, in a spiritless wandering from one historyless housing development or rental apartment to another. This unmooring is reflected in the everyday life of urban lifestyles. Junk food, mall architecture, the strip, canned entertainment, and sentimentality replace the sphere of participatory citizenship at a local level in suburban America. Planned communities and developments with deed restrictions and unspoken covenants have become a dominant American experience of neighborhood. These sites, cordoned off by freeways and strip malls, are frequently isolated spatial experiences dependent on the reversibility of the automobile. Such a fractured spatial relationship between public engagement, street life, and the neighborhood evacuates commitment to the local, leading to a state of nomadic disaffection.

Nomadic urbanity does not necessarily imply movement. Displaced and fragmentary relationships to place are part of modernity's violent logics. Physical experiences of dislocation lead to forms of political vulnerability. In the case of global migrants in varied states of illegality, their condition of nomadic urbanism is predicated on deferral, denial, and nervous attention to the formal demands of legal citizenship. Hence, the surreptitious political apathy that seeps into the everyday is hard to detect and even more difficult to counter in the age of "total information" ideologies.

Despite its attendant alienations, urban nomadism merges routes of transnational migrations on a scale unimaginable before. It is linked to multiple networks of communications, including social networking sites, cell phone technologies, and e-mail, forms that are converging physically displacing experiences. These circuits are linked through a growing need for cheap labor, despite a diminishing international culture of tolerance for immigrants and refugees. Panic is shifting migrancy to a new scale of uncertainty and fear.

In opposition to this state of nervous nomadism is the somnambulistic nomadism of American suburban life that lulls citizens into a state of political inertia fed by the delirium of television and the scale of commuting that marks everyday routine. With the automobile demarcated as the mobile home of the American lifestyle, political life has shrunk to the fragmentary encounter with the supermarket, gas station, and the megamall. The demise of engaged neighborhood politics and the sphere of the local within suburban imaginaries opens the door for the ultimately totalitarian possibilities inherent within modern urban nomadism, distracted by the individual consumption imperatives that are viewed as the full extent of freedom's promises.[8]

In coastal cities, such as New York, in particular, nomadic urbanism acquires a distinctly liquid shape as communities from different parts of the globe relate to such a city in specific ways. Outsourced work in India, cheap labor from Latin America, strategic service economies from Europe—these different relationships structure very distinctive relationships of kinship and diaspora. Monetary remittances, networked families spread across continents, and digitized domestic relationships subsisting on phone calls and Facebook connectivities enable new relationships between landscapes and peoples who are nomadic and unmoored. The distance created by large bodies of water force more inventive methods of keeping families intact, of holding lives together, and of making new kinds of futures possible for migrant subjects held hostage to economic necessities and legal vulnerabilities in countries like the United States.

In a state such as the United States, where the good citizen is the citizen who consumes the most, consumption replaces the spheres of public encounter as the structuring web of national affiliation. Consequently, nomadism becomes the seductive panacea to the increasingly abstract, dislocated, and distanced relationship between citizen and state, between citizen and neighborhood, between citizen and the city. Nomadism rears its ugly head as the terrifying condition of possibility that engenders new forms of authoritarian control, disguised in the attractive package of the free-market economy: be mobile, feel in control, but be rendered essentially powerless.

New York City is the most overexposed city in the world. Its signifying role is the claim that New York is founded on the invention of free enterprise, a colonial trading post turned global trading city. Anyone can come here and reimagine herself. This raw promise of possibility is what remains after the dreams, the arrival, the shock, the grit, the thrill, and, eventually, the calm of the city. Yes, the rules have gotten more complicated, the experiences more ephemeral. Still, what impresses one about this city are its communities of engagement. Collisions of different kinds of people within such a tight space allow for unexpected intimacies, shared travails, contingent collaborations, and tenuous friendships. Here, even for the wealthiest of New Yorkers, the density and discomforts of daily life render it difficult to remain disengaged.

Frugality and Urban Life

"Everybody sacrifices a little," Mayor Michael Bloomberg once observed. The modern metropolis is bound neither by moats nor city walls, neither by asceticism nor excess. Its perpetual proliferation of desire and restraint is a condition of immersion in a landscape of reciprocity, of what people create and what they take away from a city. In 2008, the notion of frugality made a resounding comeback on the streets of New York City. People spoke of frugality as though it were the new excess. "Be frugal" was a motto among many parents at PS 41, Greenwich Village, in 2009. The fashion and merchandising markets promoted frugality as a way for the times. This new frugality is a catchy sentiment. Its roots lie in earlier discourses on parsimonious consumption. Frugality is an international and interlocking phenomenon, grounded in earlier economic convulsions, particularly in former Second and Third World milieus. It has become, for the first time since the Great Depression, part of the American cultural fabric.

Living in New York City is a commitment, and frugality is implicitly part of its motto. New York's voraciousness is part of its lure. Flesh and bone impacts concrete, glass, steel, and speed, generating the movable landscape of pleasure and abstinence. Yet somewhere at the heart of the tension between pleasure and denial, between enjoyment and rejection of the city's myriad opportunities, is a hierarchy of notions structured around the idea of more and less. Everything in a city dweller's life is ultimately reduced to these two simple principles: more or less. Too much money, too little space. Too much distraction, too little time. Too many people, too little privacy.

In a city as diverse as New York, where everyone who enters its realms has a stake as a citizen of the city, the relationship of more to less assumes a complex set of adjustments and deferrals, best encapsulated by the term "frugality." Frugality is an economic condition and social practice. However, for an immigrant urban dweller, a nomad or sojourner, the condition of frugality becomes more than an economic necessity. It becomes an aesthetic of everyday life, a way of being in the city of one's future—an urban aesthetic that is very modern, very urban, and very New York.

New York is a system of desire, sacrifice, and enjoyment woven into the intricate economy of immigrant labor that enables the city to perpetuate its fiction as the pleasurable city, the city that never sleeps. For immigrants and the underclass of New York, pleasure is experienced through an elaborate network of checks and balances, both psychic and material. This ecology of need and satiation, this frugality, is a fundamental principle that revolves around want, deferral, and learning not to want.

Foreign banks noticeably populate Chinatown, in Manhattan. The idea of saving money, or sending all one's earnings back home to China or Bangladesh, precedes the immediacy of pleasure brought on by spending. For immigrants laboring at the peripheries of urban markets, the luxury of being in the metropole is often burdened by unseen kinship systems extending halfway across the globe.

The complexity of urban living unites the principles of frugality and perpetual desire. Living in a city involves sacrifice, with scarcity an operating principle determining the value placed on the object of desire. Scarcity always conceals the desire to acquire, to consume.

Frugality as Urban Aesthetic

Frugality in the twentieth century was mobilized within the First, Second, and Third World arenas as a structuring concept that simultaneously expressed the different economies of expenditure across the global divide. Urban expressions of state ideologies of frugality in the West and in decolonizing countries produced particular experiences of twentieth-century modernity. Socialist cities, such as Dar es Salaam and Cairo, in Africa, and Sarajevo, in the former Yugoslavia, are linked by the stark international style of architecture found in Marxist cities in the Third and Second World imaginaries. Often marked by postwar socialist housing, signs of urban decay, eroding residential structures, and nineteenth-century graciousness converted into twentieth-century overcrowded housing, these cities bear an aura of frugality through the public staging of urban neglect.

Frugality, in the postcolonial era of the 1960s, was a modernist project. It emerged as a result of the three-world system and a bifocal logic of excess or scarcity, capitalism or socialism. Operating on this transnational logic, frugality manifested the negative side of expenditure, the underbelly of conspicuous consumption. Frugality and democracy formed linked rhetorics of the modern city, as a right to the city and the rights of the individual blurred the boundaries between needs and rights. Less space, more housing. Less horizontal expansion, more verticality. Less spending, more thrift.

The history of frugality in Western market economies in the mid-twentieth century suggests a variety of other relations between abstinence, enjoyment, and expenditure. In the United States, frugality links with a notion of restraint, old wealth, and measured spending that denotes expendable capital. Here, the frugal city—the city that combines mass housing with minimal expenditure—redefined Bauhaus style as proletarian. In communist Europe, the minimalism of the Bauhaus aesthetic merged with the frugality of socialist policies and generated a new postwar framework of uniform housing that was rational, devoid of character, and driven by productivity. The resulting mushrooming of mass housing in the form of microcities, familiar in the urban United States as public housing complexes and in Third World cities as the sign of modernization, created new conceptions of frugality in relation to economic need and aesthetic minimalism.

During the 1970s, the ecological and environmental movements—connected expressions of frugality—invoked images of a depleted earth. Vegetarianism and the popularity of books such as Frances Moore Lappé's *Diet for a Small Planet* (1973) marked a self-conscious move toward a selective frugality through the exclusion of meat, but not, ironically, a prohibition on automobiles. The vegetarian represented a rejection of a certain kind of meat-eating culture of excess in the United States, while maintaining contradictory relations to consumption through dependency on the oil economy and car culture.

Frugality had a wider currency as a socialist aesthetic and state ideology during the first three quarters of the twentieth century in many Third World cities that imagined themselves as "Second World" cities. The principle "less is more" shaped relationships of exchange in cities like Prague, Dar es Salaam, Belgrade, and Moscow, during the Cold War. Frugality operated under a different logic in American cities, where ecological and environmental arguments against consumption gave rise to lifestyle choices such as vegetarianism and recycling. This subsequently became commodified in the 1980s and 1990s. Sentiments of less is more, linking frugality to urban concerns, were also mobilized in New York City during the 1970s, during the energy crisis and the blackout of the city, in 1975.

In a Second World context, such as Tanzanian socialism, frugality was a means of addressing a postindependence transition crisis. The postcolonial city became a space of self-invention. In Dar es Salaam, as in Eastern European cities like Riga or the former East Berlin, the physical layout of the capital city mirrored the ideology of the state. Socialist housing, public monuments, and statues commemorating the socialist state and its citizenry delineated the main transportation arteries and junctures linking the city to its extremities of socialist flats and communal developments.

Frugality as a state policy of self-reliance, or "Ujamaa," articulated in the Arusha Declaration of 1967, in Tanzania, aimed at a redistribution of resources via the restructuring of education, physical culture, and public housing. Through the strategic use of public spectacles like youth camps, national holidays, national parades, and public dance performances on the streets of the city, frugality became institutionalized as a human-scale proposal essential to the recuperation of a postcolonial economy.

A Performance of Duty

Modern cities embody elaborate forms of frugality. They are economic, visual, tactile, and visceral. The frugal is an obsessive condition. It is an insatiable desire that vicariously distorts urban need. In the rhetoric of policy, frugality was a concept associated with Chairman Mao's experiments with self-subsistence and mass education in Yenan Province, China. In 1960s Tanzania, the first postcolonial Tanzanian head of state, Julius Nyerere, theorized a policy of economic restraint as a way to cope with the economic stranglehold of Western dependency and debt. In the African context, frugality emerged as an economic policy, a public commitment, and a social issue, during the first decade of postcolonial Tanzania—both a social philosophy of enforced daily practice and a necessary way of life, represented by spare shop windows and few luxury goods.

With the official demise of the three-world logic, in 1989, the concept of a postmodern frugality has gained informal legitimacy. Articulated in the West through environmentalism, ecological movements, and rhetorics of anticonsumption, Western frugality has come full circle to an earlier moment of anticonsumption, one laid out in the pioneering sociologist Max Weber's analysis of the Protestant ethic and filtered through the excessive 1980s. Postmodern frugality is a layering of histories of anticonsumption, both capitalist and socialist. In his treatise on Protestant ethics, Weber advocates austerity as a way of socially managing the desires of the unruly body. Austerity becomes an expression of Calvinist rationalism. It is a response to the temptation that wealth might present in the guise of "sinful enjoyment of life" and "living merrily and without care." But as a performance of duty in a calling, austerity is not only morally permissible, but actually enjoined.[9]

Weber points out that the emphasis placed on the ascetic life was bound to the development of a capitalistic social order. A revulsion against the ways of the flesh in early Puritan society supported the rational acquisition of wealth as much as it abhorred irrational expenditure and consumption. This intrinsic contradiction within capitalist productive ethics fuels a fundamental urban tension. In Weber's terms, it is the contradictory site of capital accumulation through the ascetic compulsion to save, combined with the temptation of wealth itself as desire for worldly enjoyment.[10] Such an emphasis on "the ascetic importance of a fixed calling" provided the justification for a utilitarian way of life, one in keeping with Weber's modern, specialized division of labor.[11] If asceticism, as Weber suggests, suited the mechanisms of capital, then enjoyment detracted from productivity, the goal of capital. Asceticism was harnessed to notions of the civic in the Puritan ethic, forging conceptions of the Puritan urban that were decidedly against consumption and pleasure. The city was a place of industry and productivity, yes, but it was also a licentious threat to the capitalist enterprise because it encouraged hedonism.

While Puritan asceticism was a social aesthetic linked to the capitalist enterprise—and by extension, to the founding of the capitalist city—its artistic equivalents of minimalism, frugality, and modernist simplicity merged discretion with extravagance in the American city. This basic contradiction—between excess and austerity, enjoyment and restraint, corpulence and thinness—underwrote the logic of modernity, operating in temporal-spatial disjunctions of the city.

In Weber's theorizing of modern urbanity, the medieval city represents a perfect fusion of fortress and market, an enclave of militantly competent citizens.[12] The geography of the medieval city organized labor and leisure in structurally distinct social categories. Social identities and urban possibilities were closely interwoven. Guilds, organized by trade, such as merchants and artisans, coexisted with other autonomous civic institutions for urbanites. At the core of this emergent social restructuring was an elaborate network of urban rights—the legal, cultural, and sociopolitical network through which civic citizenship in the medieval city operated. These systems of urban rights made available new ideas of civic unification. For Weber, a city can develop only under special conditions and has its own internal logic. It is a self-contained system of laws and spatial regimes. Its network of social relations is an aggregate of economic, religious, and political institutions. In conceptualizing the crucial elements of a city, Weber argues that a settlement has to display the following: a fortification, a market, a court of its own, autonomous law, a related form of association, and at least partial autonomy and autocephaly.[13]

The shift in geography from the medieval to modern city was also a shift in human flows. By the nineteenth century, changing notions of speed, distance, and temporality had rapidly introduced a mercantile modernity to the European city. From Henri Pirenne's medieval cities, with their nascent notions of municipality and communal organization, to Weber's patrician and plebian cities, with their evolving sense of civic participation, the geography of the city redefined how people conducted themselves within urban space.[14]

The nineteenth-century city saw the slow dissolution of the strict boundaries of the geography of place, as the marketplace, court, and fortress expanded their mercantile and civic interests in multiple ways. Industrial cities, such as Dickens's London or colonial Madras, embodied this rapidly diminishing distance between fortress and marketplace. In Manhattan, the Financial District's engulfing of Castle Clinton signified this shift in emphasis from security to finance.

The Stranger, Frugality, and Pleasure

The pleasures of the street, as Charles Baudelaire's flâneur demonstrates, and as the German sociologist Georg Simmel's figure of the stranger independently proffers, are a productivity of another sort. The street in their writings contributes to the wealth of nonascetic urban mental life, at once profoundly modern and in tension with the logic of its Puritan counterpart. As both Baudelaire and Simmel observe, the exponential expansion of populations in nineteenth-century Paris and Berlin demanded new ways of perceiving reality. Changing, and often conflicting, conceptions of excess and restraint shaped ideas of enjoyment in the emerging modern European city.

For Baudelaire, "every age had its own gait, glance and gesture."[15] And the urban body of the nineteenth century was an amalgamation of the frugal and the excessive, of spiritual and material reality. Baudelaire's meditation on the flâneur, who strolls the city with the express intent of absorbing its varieties, elaborates upon the imaginative spheres of enactment of the urban body. He observes that corporeality and optical perception must be imaginatively fused in order to comprehend the sensorial regimes that constitute the modern moving body in the street. The three-dimensional scoring of living "style" through spatial as well as temporal aspects, such as fashion and memory, offer a barometer of modernity that "time imprints on our sensations."[16] Modernity's body is at once an architectural construct and a moving display, Baudelaire observes. It sets itself against a changing, urban mise-en-scène. The moving body becomes a densely layered event of flesh, musculature, and bone, upon which an even more elaborate set of structures is imposed, in the way of fashion, gestures, and mannerisms, offering glimpses into modes of self-invention and self-interpretation. The flâneur is of the crowd, immersed in it, but at the same time individuated in the sense of the meditative self. The modernity of the nineteenth-century flâneur lies in the self-reflexive realization of everyday life through the spatiality of the body.

Implicit in the Baudelairean vision of the flâneur, however, is the sense of cultural sameness structured into the feeling of solitude caused by urban modernity and its resultant distancing. The Baudelairean flâneur is a stranger in a city he knows well, Paris in the nineteenth century, as much as his strangeness is caused by the density of migration to the city he inhabits with such intensity. He is also a cosmopolitan traveler, always with an eye on capturing scenes of everyday life on the world stage. Scenes from the "garrisons of the Cape Colony and the cantonments of India" give one a glimpse of the array of experiences the true flâneur imbibes of the imperial world. In the flâneur's world, strangeness has textures and depths as varied as the global nineteenth-century market: "And now we are at Schumla, enjoying the hospitality of Omer Pasha—Turkish hospitality, pipes and coffee."[17]

As in Baudelaire, the modern individual in the city is the mediating point for Georg Simmel, the German sociologist. Individuals and urban social relations hinge upon the category of exchange, which always involves some tinge of sacrifice, whether it is through the deferral of desire or a burst of extravagance that will require payment in the long term. For Simmel, however, the crucible of modern individualism, the city, was also the scene of unexpected encounters with strangers. The stranger was at once a central aspect and sign of the city's chaos. The migrant, the immigrant, and the traveler all embodied a realignment of fundamental social relations that could alter notions of culture and community in unforeseen ways and destroy existing social relations articulated around the familiar and the insular.

Writing in the same city, Berlin, at the same time, the sociologist Ferdinand Tönnies similarly echoes an anxiety around the nineteenth-century expansion of *Gesellschaft*, or society, at the expense of *Gemeinschaft*, or community, as urbanization and modern individualism reshaped the relationship of people to their rapidly changing environments and, in turn, their sense of context and experience of place. Nostalgic for supposedly binding forms of community embedded in a fast-disappearing folk culture, Tönnies articulates a rising sense of alienation at the end of the nineteenth century. For Tönnies, the psychic disruptions wrought by industrialization had introduced indifference as the scourge of modern urban life. "But our souls, our feelings," he laments, "are indifferent to the great mass of people, not only to those who are unknown to us, the strangers, but also to those whom we know reasonably well."[18] This specter of indifference, whose sign was that of the stranger and the immigrant, for Simmel as well as for Tönnies, is crucial to the narrative of what makes the city possible. And it is one of the axial points around which ideas of the civic continue to be articulated.

The Manhattan Way of Life

Cities conjoin frugality, perpetual desire, and the complexity of urban living. They involve sacrifice. Financial scarcity, in particular, shapes urban desire. In 2009, this catastrophic combination threatened the very way of life in New York City, as regular visitors to soup kitchens surged, unemployment statistics escalated, and New York's populations braced for more bad news on the job front. People who once earned comfortable middle-class incomes found themselves dangerously on the brink of bankruptcy. Two-income households careened under the stress of both householders losing their jobs. Neighborhoods that ballooned in the days of subprime loan mortgages had fallen vacant, bringing chaos to vulnerable populations. For immigrant urban dwellers, or those located in uncertain jobs, the challenges of living in New York were further exacerbated by the stress of families left behind, sometimes halfway across the globe. In such a climate of restraint, frugality had temporarily established itself as a way of being in the modern metropolis. The Manhattan way of life was a formidable yet hopeful array of deferred desires and innocuous pleasures.

CHAPTER 6

nyerere, the dalai lama, gandhi | *cultures of frugality*

On May 11, 1998, Julius Nyerere, who had been the first president of Tanzania, gave one of his last talks in New York City, at a venue south of Washington Square Park. His death, a year later, marked the passing of a key phase in the history of modern metropolitanism—the dramatic invention of new selves in postcolonial cities around the world. As he walked into King Juan Carlos II Center, at New York University, wearing his characteristic Tanzanian blue suit, Nyerere's corporeality represented a convergence of this myth, an icon of African history, and a cultural moment—the end of an era of utopian imaginings.

In Dar es Salaam, where I came of age in the 1960s, urban postcolonial self-inventions were unconsciously impacted by the glamor of American urbanism, received as a new international commodity through the images of Shaft and Super-Fly, Foxy Brown and James Brown. As a multiply migrated Tanzanian-now-New Yorker without an easy narrative to connect the Second World experience of my childhood to the intensified reality of New York City, observing the poignant matrix of postmodern New York alongside the icon of my own African socialist youth, and the heterogeneity of the packed New York auditorium, made tactile the dynamism of historical junctures. Nyerere's physical presence south of Washington Square Park shrank the distance between the socialist Dar es Salaam of my teen years in the 1970s and an excessive millennial New York, just before it all came apart in 2001.

The montage effect of a living icon of African socialism appearing in the landscape of downtown Manhattan highlighted both the space of the encounter, that of millennial Manhattan, and the historical actualities linking the African socialism of the 1970s to broader global movements of scarcity and frugality impinging on the world today through the discourses of ecology and environmentalism. Suddenly, my dramatic and seemingly disconnected experience of Tanzanian Ujamaa, or self-reliance, in the port city of Dar es Salaam during the 1960s and 1970s, was brought into conversation with New York City as a port city whose own current interests in public space and environmental sustainability echoed one of the most powerful tenets of African socialism from an earlier era: frugality as an aesthetic of urban life.

Remembering Ujamaa in Manhattan

It was a gorgeous day in late October of 2007, in Washington Square Park. The park was filled with life. Lounging city dwellers basked by the soon-to-be-demolished Washington Square Park fountain in all its spectacular off-kilter, decaying glory. Colorful vendors filled the north side of the park. Battalions of children ran, screamed, laughed, and cried. Some waded through the fountain's spraying water.

Amid such distracting mayhem, something beautiful was happening. The children of Public School 41, in Manhattan, were planting bulbs for flowers, as were students from other schools across New York City's public parks. My daughter planted her group of bulbs excitedly, along with nearly seventy other children, as part of the citywide initiative under Mayor Bloomberg to green New York City. The plan to plant ten thousand more trees was an effort to counter the warming of the planet through the expansion of green space. The act of planting is a special moment, an intimate space between a child of the city and the city itself. It is a deeply restorative act of metropolitan investment that requires digging into an earth usually covered by concrete.

I was suddenly reminded of my own intimate moment of planting a tree in another era, in another city, as part of a different utopian project, and yet not so different a project within the context of global politics. It was my memory of planting a tree along the sidewalks of the major boulevard, Independence Avenue, as a primary school child in Dar es Salaam in 1969, as part of the national initiative of Ujamaa. Once a city child, I could recall the excitement of digging into the earth in the midst of a busy avenue, to plant trees for public enjoyment. At the time it seemed an unclear project, an entertaining waste of my school time. But now, repeating the act in the shadow of Stanford White's Washington Square Arch, I grasped the utopian imperatives of children planting trees in public parks. It was a tactile act of participatory civic engagement—an action of public enjoyment and a permanent link to the shaping of a city's future.

In that moment of digging into the mud, the utopian project of Ujamaa merged with the needs of the present time, the need for direct action by greening New York City's public spaces. Using a *jembe*, or spade, some bulbs, and a child's hand, the city of New York in 2007 was stepping into a new commitment with the landscape of the city. It was a commitment that was slow in coming to a full realization, but one that has made many inroads toward the civic reeducation of urban dwellers.

During the Arusha Spirit of 1967, in socialist Tanzania, a season of marching gripped the nation's imagination. Many walked from the remote regions of the country to Dar es Salaam, the state capital at the time, to express their support for the Arusha Declaration. Toward the ceremonial recognition of this monumental event in the history of modern Tanzania, the then-president Julius Nyerere took the 134-mile trek from Dar es Salaam to Mwanza, thus marking the boundaries of the self-inventing nation. The march began a season of change, a dynamic renegotiation of the urban and the rural, city and country, on new terms, that of the postcolonial nation-state and its attendant notion of what Tanzanian citizenship might mean. The following anecdote, describing the tail end of the long walk, captures the moment: "At last the President called a ten-minute halt. 'Well fellows, we've done two and a half hours.' Hashim Mbita emptied the sand from his shoes. Nyerere noticed the shoes worn by Joseph Namata, now the commissioner of Mwanza region—brown suede with pointed toes. 'Ah, Manhattan,' Nyerere said, feigning scorn. 'Not so good.'"[1]

A great proponent of walking, Nyerere embraced Mahatma Gandhi's belief in walking to create a sense of nationness never possible before. Walking, in Gandhian political strategy, is a form of nonviolent action that leads to political consciousness. Nyerere took his cue from Gandhi's own political maneuver: to walk across colonial India in the famous Dandi March, which led to the largest spontaneous mass mobilization of peasantry across India, and eventually to independence. This strategy generated corporeal knowledge of national identity. For Nyerere, this visceral mapping of the new nation was necessary to create a new kind of psychogeography for the new citizen of the postcolonial state.

The "Manhattan Way of Life"

To further the utopian goal of creating new urban African subjects, Julius Nyerere gave many speeches to primary school students growing up in the Africanizing capital city of Dar es Salaam, with its Swahili medium schools. One of his goals was to cultivate an aesthetic of urban frugality. Frugality, Nyerere emphasized, was an important sensibility of modern life.

The good "Mwalimu," or teacher, as Nyerere preferred to be called, frequently reminded Tanzanians that they could not afford the "Manhattan way of life," a phrase he used often to emphasize the excessive effects of capitalism, largely embodied for us schoolchildren in the shape of red Corvettes, tight miniskirts, platform shoes, the Afro, and Coca-Cola. Manhattan epitomized the apotheosis of American modernity, from which Tanzanian urban life would have to distance itself. This Manhattan way of life conjured up vague, popular notions of high living, hyped-up fashion, oversized cars, and degenerate lifestyles. Overconsumption, implied by the phrase "Manhattan way of life," would not work for an African urban modernity predicated on self-reliance, Nyerere argued.

Instead, as Nyerere put forth in his economic manifesto of 1967, the Arusha Declaration, African modernity demanded a more viable and locally sustainable mode of development, grounded in frugality as a daily practice. Nyerere's advocacy of a culture of less consumption had effects on the city of Dar es Salaam, with no driving on Sundays and the injunction to drive less and walk more during the height of the oil crisis in the 1970s.

The frugality advocated by the Arusha Declaration of 1967 posed an antiurbanism that reflected the tensions within Second World African modernization initiatives: the struggle between urban development and agrarian modernization. In the interests of the democratization of the rural peasantry, the declaration promoted self-reliance as a way of accelerating rural modernization. What emerged, however, was a fundamental contradiction under international socialism. Democratization implied an implicit tension between the urban masses and proletarian movements in rural areas. The effect of new policies to democratize a primarily rural population led to the growth of migration to cities. An influx of expatriates during the rapid nationalization of the country's industries created a two-tier economy: a local culture that was Africanizing and nationalizing, and an expatriate culture that took on a sojourner relationship to the emerging urban landscape of peasants turned city dwellers.

In counterpoint to this notion of frugality, a Manhattan-style urbanism was invoked by Nyerere's use of the phrase "Manhattan way of life." This expression, embedded in my mind as a schoolgirl in the Dar es Salaam of socialist proleterianization, conjured up more an image of Los Angeles-meets-*Superfly* than any comprehension of a New York lifestyle that was predicated on a walking city whose rhythms were much closer to those of downtown Dar es Salaam. Furthermore, the phrase appeared to primary school students to be a damning indictment of our failure to be devout socialists. Yet, in retrospect, New York then and now shares many of the same concerns for democratization, sustainable development, and the impacting dialectic between scarcity and excess, frugality and overconsumption, that shaped the tone of public discourse unfolding around notions of metropolitan identity in Dar es Salaam, Tanzania, during the 1960s and 1970s.

Contrary to Nyerere's use of Manhattan as a foil for socialist imaginings, Manhattan's way of life was much closer to the logic of urban circulation in Dar es Salaam than any other Tanzanian city at the time. The relationship of center to periphery, which structures New York's relationship to other cities around the world, resonates much more with a Third World capitalist city in a Second World political imaginary of international socialism than it would in a rural town such as Arusha or Morogoro. Dar es Salaam's geopolitical location as a sixteenth-century colonial shipping port, was transformed into a thriving postcolonial capital city of a newly independent state juggling the delicate balance of power between American Fordist efficiency, Soviet models of modernization, and Chinese proletarianization. This city has more in common with New York, which experimented with communes, utopian communities, forms of socialist organizing, and attempted to lift up the languishing immigrants of the Lower East Side from formerly totalitarian regimes, than rural Tanzania has in common with urban Dar es Salaam.

After his powerful speech in New York City in 1998, in which he called for a more concerted African leadership toward economic parity in Africa, I walked up to the grand old man. I had shaken his hand as a child at St. Joseph's church in Dar es Salaam many times and had shared a desk with his daughter Rosemary while in primary school. Now, here at Juan Carlos Center, next to Washington Square Park, in New York City, I thought I should say hello. I mentioned I had written a book about Ujamaa and a chapter about him.

"I hope you said some good things about me," Nyerere said wryly as he held my gaze.

Nyerere's comment moved me in some way. Ever the visionary, with a lifetime commitment to justice and equality, his experiments in social engineering received mixed assessments from economists and politicians alike. What remains in the texts of many of his speeches, however, is Nyerere's full awareness of the challenges of urbanization. He had been learning from Manhattan all along, emphasizing that Tanzania had to imagine a different kind of urban future than the one postulated by American consumption. That future would be more utopian than viable, as the oil crisis hit Tanzanian development in the 1970s, ushering in an era of urban frugality expressive of a global recession that affected New York City's landscape during that period as well.

Dalai Lama, SoHo Kitsch

On a cold January day in 2008 a man stood naked on Seventeenth Street and Seventh Avenue, in Manhattan, buried up to his neck in glistening blocks of ice for a period of one hour and twelve minutes. Crowds gathered curiously to watch the Dutchman Wim Hof break his own world record of being buried in ice. It is an unspectacular scene, just another performer on a street corner doing something extraordinary—in this case an act of endurance outside the Rubin Museum of Art, the repository of Tibetan Buddhist Art in New York City.

Hof's scientifically baffling ability to survive in subzero temperatures for great lengths of time connects to an underlying thread of life in New York City: endurance, the ability to survive extreme situations. Hof is a proponent of Tumo meditation, a form of deep Tibetan meditation that predates the incursion into the region of Buddhist forms of bodily practice. The juxtaposition of Hof's feat of extraordinary yogic control with the urban backdrop of Manhattan connects with a deep-seated preoccupation in New York with bodily forms of preservation within urban life. Evidence of New Yorkers' lively attraction to alternative forms of bodily control and lifestyle philosophies takes many forms, from the statue of Confucius at Confucius Plaza in Chinatown, built in 1976, to the almost annual visits of the Dalai Lama.

The city's growing interest in alternative medicine, and alternative therapies such as ayurveda, Thai bodywork, Chinese Qi Gong street massage, and Chinese medicine, has been institutionally grounded in the Rubin Museum's programming, which exposes New York to the more esoteric Tibetan and Indian yogic traditions as a possible way of urban life. The

Dalai Lama talked to packed audiences at Radio City Music Hall in 2008 and 2007, and in Central Park in 2002, adding to the ongoing conversations among New Yorkers about alternative urban lifestyles and attendant body regimes for a sensuous urbanism.

The Dalai Lama in Manhattan

Wielding placards protesting the Chinese occupation of Tibet and its clamping down on Tibetans in Tibet, Himalayan communities in New York took hold of the west side of Manhattan's most public corner to demonstrate against Chinese oppression: "Long Live Dalai Lama." "Long Live Panchen Lama." "Free Tibet." "One World One Dream." "Boycott Beijing." Against the backdrop of the impending Beijing Olympic Games, in August 2008, the Hudson River Park bike path at the junction of Forty-Second Street and the West Side Highway, across from the Chinese Consulate, became a temporary staging ground for daily protests. Over many months they gathered, from the spring and summer of 2008, through the hard winter, into the summer of 2009. By August 24, 2008, the protests had become more vocal, more emphatic, more colorful.

On July 17, 2008, the Dalai Lama's visit to Radio City Music Hall for a fundraiser triggered unprecedented unrest and skirmishes along Fiftieth Street and the Avenue of the Americas. The disruptions were a result of tense encounters between the Dalai Lama's devotees and protestors from a rival Buddhist sect, the Western Shugden Society. Spring of 2008 saw an escalation of the visual and written invocation of the Dalai Lama all over New York City. The iconic figure of the Dalai Lama as a Manhattan presence had taken hold of the city again.

The Dalai Lama has had a very public presence on the streets of Manhattan for nearly two decades. During the spring of 1998, Apple Computer's ad featuring the Dalai Lama went up as an enormous street poster at the junction of Houston Street and West Broadway. At the time, SoHo was Apple's first location for a store in New York. Part of Apple's worldwide advertising campaign, the poster—alongside another of Amelia Earhart—marked the border of the art-focused district of SoHo, suggesting the spirit of technology and creativity proffered by the neighborhood and its residents. The tagline of the ad urges viewers to "think different." The popular circulation of the Dalai Lama as a commodity, alongside other world icons of science and revolution—including Albert Einstein, Pablo Picasso, and Mahatma

Gandhi—marketed kitsch, New Age consumption, and urban spiritualism as much as new technologies.

Historically framed at the crossroads of desire and political exile, filtered through the exotic notion of Tibetan Buddhism as mystical and timeless, the benevolent smile of the Dalai Lama at the border of Greenwich Village and SoHo marked the sentient formlessness through which the body operates in the city. Urban modernity is nurtured by the inner secrets of the mystic East. No postcolonial reflexivity here. No pausing before the possible belittling of a religious icon emblematic of territorial disfranchisement. Instead, the moment, framed by Apple's internationalist embrace of revolutionary world figures, merely reiterates an ahistorical quotation: free-floating mystic Tibet freshly framed by the Hollywood film *Kundun* and Brad Pitt. The East is back, retooled with new technologies, to offer the secrets of ayurveda and the Himalayas without compromising the relentlessly individualistic logic of capital's euphoric self-obsession. The Dalai Lama merges with the Hindu mysticism of the chain beauty salon Aveda, also at the corner of Houston and West Broadway. His image is reduced to that of another performer in a city full of celebrity, despite his symbolic prominence in international politics.

Writing about the visual pleasure of looking at cities, Kevin Lynch, the urban planner and author, remarks: "At every instant, there is more than the eye can see, more than the ear can hear, a setting or a view waiting to be explored. Nothing is experienced by itself, but always in relation to its surroundings, the sequence of events leading up to it, the memory of past experiences."[2] The poster of the Dalai Lama brings into play the cumulative effect of the sensorial in the city: dense layers of commoditization and the spiritual.

How does the sentient individual experience such a merging transaction of meanings and landscapes? What subtle interplay of context and desire is set in motion through the juncture of an image of a Tibetan monk, who is also a stateless sovereign synonymous with a metaphorical Tibet, and that of SoHo, whose own economy of affluent sanctuaries, craft stores, and corporate monasticism is combined with the rarified aura of new media technologies? The interiority of SoHo, at once esoteric and renunciatory, is signaled in this display of consumer and corporate advertising. Encountering the Dalai Lama amid the cityscape triggers sensations of other orders of being, ways of living yet to be mastered.

The choice of SoHo, in 1998, as the site for Apple's grammatically questionable imperative to "think different" is a strategic one of branding.[3] Historically, SoHo has been the outpost of innovative lifestyles created by artists' reclamations of industrial spaces during the 1970s. SoHo's population of alternative art enclaves and experimental galleries, from the 1970s to the excessive 1980s, brought along with them an interest in New Age life choices. Only in SoHo could the techno-hip spiritualism popularized by the scientifically committed Dalai Lama meet the technological feminism of the first female pilot to make a transatlantic flight, Amelia Earhart. The two images emblematized a mythic emergence of SoHo as corporate ashram.

By the late 1990s, SoHo's labyrinths of decadent slumming, overconsumption, and flamboyant meditativeness had emptied it of its edgy loft-living aura. Still, the rarified atmosphere in which to "think different" amid anorexic models of all hues brushing shoulders with high-fashion CEOs on their lunch breaks, architects, and designers, captured the lure of SoHo. The image of the Dalai Lama under the corporate logo of Apple, at the corner of West Broadway and Houston Street, served as a tranquil reminder of the digital revolution more than the territorial disempowerment and exile narrativized by the Dalai Lama himself.

In the Paris of the nineteenth century, Baudelaire saw all the juxtapositions of fantasy and reality cumulatively producing new sensual experiences of the city. For Baudelaire, the city is never knowable. It is elusive and seductive, corporeal and flat-spaced. Particular locales in the city remind one of a particular image now gone, or an event long forgotten. Yet, as the poster of the Dalai Lama demonstrates, such sites of the city sometimes become imprinted with a specific aura that has vanished, but still captures an aspect of the city's mental imagining of a specific moment. Such imaginings may be individual, as in particular interpretations or experiences of the city, or public, as in the performances of street vendors negotiating their relationship to the city. Yet such elusive mental images, though fractured and nebulous, cumulatively produce a mental image of the city that is individually specific, local, and even culturally particular.

In writings about New York, the city is invoked as architectural space, as landscape, as a conglomeration of buildings, as a playground for urban planners and designers, as sociological category, but its tactile and particular ways of knowing always remain beyond the realm of the empirical.[4] Its unexpected juxtapositions of history, memory, and the real, as in the rare conjunction of the Apple poster and the Dalai Lama's very visible presence during his visit to New York in May 1998, and in numerous visits over the next decade, generate cumulative questions around historicity and embodiment, periodicity and ways of knowing.

Lama Rama

Located at a bustling junction of the city, prominently displayed atop the streetlights, the image of the Dalai Lama at the corner of Houston Street and West Broadway evoked mixed feelings of pop reverence, postindustrial serenity, and political ambivalence. Against this public, iconic display of technological futurity and high fashion, a visual embodiment of SoHo, the actual Dalai Lama's presence acquired a less impressive stature. "He is not as imposing as he looks in the poster," one New Yorker said after attending a lecture by him. "The Dalai Lama is a New York fetish," pronounced another acquaintance.

The explosion of lama frenzy during the middle of May 1998 worked at different levels and captured an aspect of the metropolitan good life that combines international human rights with high fashion. A penchant arose among high society for a lama at every function. The elevation of Tibetan lamas to the realm of primitivist kitsch was poignantly enacted in the Dalai Lama's attendance as visiting rock star at widely diverse events, from theological world meetings held in prominent locales to protests by the New York Chinese community at the Temple of Understanding, an interfaith organization based in New York City that is committed to human rights, and in which the Dalai Lama plays a prominent role.

The image of the Dalai Lama was taken down at the end of May 1998 in response to protests by the Chinese, who were angered by the attention he had received in Hong Kong and New York. The Dalai Lama's visit to New York in the capacity of a civic authority raising support for his cause was relegated to the sidelines as opposition to his civic authority was staged across the city, accentuated by the increasing prominence of the Nechung Oracle as a contestatory presence in the diasporic struggle for Tibetan sovereignty.

The fascination with monks within city life is as old as the idea of the city itself. Tropes such as the friar and the monk in classic travelogues like *The Canterbury Tales* document the importance of the traveling holy man, whose degree of asceticism often contrasts with the corpulent profile he bears. These caricatures fill the annals of early urban migrations.

The medieval English monk, for instance, was a fictive assemblage of greed, bonhomie, and generosity. Largely a homosocial creature accustomed to city life within the medieval church, fort, or street, the male monk bore comic connotations of excess and denial, the Spartan robes of the Franciscan friar or Capuchin monk contradicting the power and privilege exercised by the Church within the city.

The French, Rabelaisian monk of the Middle Ages was a loose-living fellow, at once scholar and peasant, loquacious and parodic. He was widely traveled, as the ways of the cloth are migrant, and acquainted with the customs of many cities, either through conversation or personal experience. Earthy, sensual, and hardy, the Rabelaisian monk is an eloquent boozer whose relation to both God and Devil are equanimous. Fundamentally social and urban, the Rabelaisian monk is a full-hearted consumer of worldly goods, a scientific connoisseur of excess and frugality. He combines the fastidious palette of Benedictine simplicity with the crude fare of Franciscan proletarianism. Simultaneously without property and claiming the world as his stage, the Rabelaisian monk stares enjoyment in the eye and takes its fleshy comportment head on.[5]

The renunciation of the urban invokes the landscape of the hermit and the sage. The hermit's reclusiveness is combined with a deeper understanding of nature, achieved largely through a rejection of the lures of the city. Unlike the urban monk, the sage takes refuge in the wilderness. Sagacity demands deprivation, isolation, abstinence. The resulting endurance leads to a profound comprehension of life's worth. Such a project would be thwarted by the distractions of a city. Indeed, the city could derail one from realizing knowledge born of nature. Such knowledge, embodied by the sage, is diametrically opposed to life in the city, yet is crucial to understanding ways of contending with the excesses of the urban, as staged by Wim Hof's performance of extreme endurance.

Tropes of the hermit and sage were popularly contrasted with that of the mendicant, another popular urban type in the medieval city. In India, the mendicant projected an aura of saintliness and abstinence. He was sworn to a life of begging, and he bore the social role of inculcating self-restraint by example into the lives of urban decadents. The state of mendicancy could be arrived at through a number of scenarios. One might be born a mendicant, or recognize the depravity of urban life and devote oneself to a life of simplicity, or fall from grace and choose the life of the mendicant, living at the mercy of the street, in search of new direction.

The narrative of the mendicant always invokes a former life of decadence and excess in the city, which the worldly soul eventually rejects for a way of life in opposition to the fleshiness of urban living. In this scenario, the mendicant is a social monk whose torment—that of self-imposed perpetual exile through travel or vagabondage—is also his means of subsistence. He is dependent on the generosity of overextended urban dwellers.

The social production of the Dalai Lama, however, presents a unique sovereignty unavailable to mortals and consequently worthy of an unparalleled distinction: the face of SoHo. He is nominated at birth as sage and monarch, steering the faithful and the ignorant away from the vicious cycle of want created by the city. He is at once humble burgher and sacred monarch in exile, forced to a life of political mendicancy, the way of the postimperial, stateless, holy sovereign. The poignant combination of technological holy man and stateless sovereign encapsulated by the patrician face of the Dalai Lama is offset by the dense mobility of bridge-and-tunnel traffic and the transient populations of touristic SoHo.

Movement of the Unfree Spirit

The historical production of the Tibetan monk as transcendental spirit has its tangents in various forms of Orientalism rooted in Western nostalgia for the unknown East. The aura of mystic Tibet lingers, minus the Chinese plastic paraphernalia and bureaucratized urban culture of cheap colorful commodities made in China and sold in Lhasa. Environmentally uncontaminated Tibet—a misguided illusion that pans out of view the plastic bags, Coke cans, candy wrappers, floating newspapers, and centralized imports—beckons as the planet furiously melts.

The political geography of Tibet has consequently been anthropomorphized into the face of the Dalai Lama. Images of chanting monks, contrasting with the impressive backdrop of Lhasa's Potala Palace, frame the complex politics surrounding issues of sovereignty and governance for Tibet as a nation. However, it is this same complexity that applies to the kitschification of the Tibetan monk as an urban type—stateless, unfree, yet sustained by elaborate forms of self-management that allow for the extraordinary charisma attributed to the Dalai Lama internationally.

The movement of the unfree is the archetypal movement of the twenty-first century—forced migrations, hasty departures, unannounced arrivals, refugee camp-turned-city dwellings. It is this movement of the unfree, epitomized by the benevolent face of the Dalai Lama, that appeals to the harried and enclosed New Yorker.

The Dalai Lama poster at the corner of Houston Street and West Broadway represents another moment in New York's history, when modes of body management, urban occultism, and Eastern forms of religiosity informed a particular subculture of the city's cultural elite. At the end of the nineteenth century, a group of New Yorkers would meet at the famed "lamasery" of Madame Blavatsky, a Russian émigré who arrived in the city in 1873. Fresh from a sojourn in India via London, Blavatsky established the Theosophical Society in her home at 46 Irving Place in collaboration with H. S. Olcott, a journalist from New York, and W. Q. Judge, a lawyer.

Under the influence of Madame Blavatsky, the Theosophical Society, from 1875 to the early 1900s, generated widespread enthusiasm for swamis, yogis, lamas, and nonsectarian mystical states. Blavatsky's own extended excursion into Tibetan mysticism—begun during her visit to Tibet, in disguise—attracted enormous interest among spiritualists in New York.[6] While the theosophists' inquiry into the occult was not unique, their urban context and cosmopolitan interests in science and philosophy kept their interrogations intellectual and international, while distancing them from the questions of race, class politics, and ideology that profoundly shaped their inroads into colonial territory.

The internationalization of theosophy as a movement with centers in London and Madras made the idea of an urban occultism widely available as a lifestyle choice. A culture of urban asceticism incorporating the mystical and theological traditions of Eastern religions, particularly Vedanta and the Buddhisms of Ceylon and Tibet, created a fascination with the colonized Orient, even within colonial cities like Madras and Calcutta. Indian mystics were complicit in the promotion of Brahmin ideology under the guise of mystical doctrines of a world religion, Hinduism. New theosophical sects and cults proliferated in New York and around the United States from 1875 on, culminating in the success of Indian gurus such as Vivekananda, who visited Chicago to attend the World Parliament of Religions in 1893, and J. Krishnamurti, in Ojai. Organizations like the Hollywood Ramakrishna Mission and Paramahansa Yogananda's "Yogoda" cult in Encinitas, California, were also successful.

Colonial fantasy and colonized complicity merged, creating a twentieth-century quietist imaginary that would blur the distinctions of West and East against the backdrop of nationalist struggles for self-determination. The theosophist Annie Besant, for instance, reinforced an important link between the neighborhood of Adyar, in Madras, and the theosophists in downtown New York. Besant participated in the Indian National Congress and founded the Home Rule League, in collaboration with Indian radicals. Further, Olcott financed support that led to the creation of the Indian National Congress. Despite Blavatsky's own disclaimer that the theosophists were nonsectarian, their political alignments with national culture in India suggest a partisan investment.

While many of the pseudocults that fell off the theosophical bandwagon descended into narcissistic and apolitical communities, other offshoots led to interests in experimental utopian communities and New Age body management, as well as ecological, paranormal, and scientific inquiry. G. I. Gurdjieff's Institute for the Harmonious Development of Man, established in New York in April 1924, was one such experimental community that extended the practice of theosophy in the direction of performance.[7] The popularity of F. M. Alexander's technique of constructive, conscious control of the individual by the individual, and Moshé Feldenkrais's focus on breathing, movement, and posture, also drew indirectly upon yoga and other movement techniques in a nonsectarian language of body management.[8]

The ongoing interest in New York in alternative medicine and yogic practices, represented by the Integral Yoga Institute, the Center for Holistic Health, and various centers offering ayurveda and naturopathy, continues to draw upon the theosophical legacy of the city.[9] "Spiritual Wellness" brochures stuck into crevices on subway trains invite the commuter to more esoteric urban regimens. Andrew Weil, Deepak Chopra, and other holistic centers offer a lifestyle of Eastern bodily regimes without the need to philosophize it. In this sense, the worst fears of the New York theosophists have been realized as an urban lifestyle choice. Encountering the East no longer involves reckless journeys into unknown territories of the empire. Instead, it has been defanged and become an avenue of consumption, just another commodity in the city's choice of pleasurable lifestyle services.

Gandhi Garden

A large group of antinuclear activists and South Asian antiwar activists were gathered around the statue of a sparsely clad man in Union Square in the summer of 1998. They carried signs that said, "India Must Divest" and "Nuclear Disarmament Now!" Speeches and passionate voices resounded through the park, set against the backdrop of the determined figure, frozen amid movement, striding uptown with frugal body.

The tenacious figure of Mahatma Gandhi breaks the continuity of traffic crowding the southwest corner of Union Square Park, in Manhattan. Made of black marble, the iconic leader stands poised in a resolute attitude of walking, inviting the harried passerby to pause and reflect around the sheltered Gandhi Garden.

The presence of Gandhi in Union Square Park serves as a public marker of New York City's liberal humanist tradition of tolerance and internationalism. The sculpture signals the transnational routes of New York urbanism. It embodies the determined steps of one who walked far and long in the interests of freedom and citizenship, words now imbued with contradictions.

Created in 1986 by Kantilal B. Patel, the Gandhi figure provocatively raises competing urban discourses on civic idealism, multiple publics, transnational urban bodily regimes, and the struggles for local citizenship that undergird New York urbanism. Located in historic Union Square Park, with its long-standing tradition as a center for staging unrests, protests, riots, and civil disobedience, the Gandhi statue gestures to the legacy of an earlier age of civic idealism that shaped the public parks and monuments of New York at the turn of the twentieth century. It visually marks the imagined linkages between an increasingly globalizing urban space and the transnationally grounded urban vernaculars emerging in cities such as New York, at once provincial yet metropolitan.

The mnemonic evocativeness of the Gandhi statue exceeds its markedly civic intentions, whose roots lie in the pragmatic efforts, from 1890 to 1930, to promote public sculpture in the interest of generating civic harmony. The interest in public monuments during the early part of the twentieth century in New York City was catalyzed by increased immigration from Europe and migration from the American South and Midwest to New York. An increasingly heterogeneous and international local population raised the question of how the city might begin a public discourse on creating shared values of urban belonging. Public monuments to such figures as Giuseppe Garibaldi and José Martí, and the Japanese stone lantern or *ishi toro* surrounded by cherry trees in Sakura Park, gained the support of the city's elites, financiers, and administrators as mechanisms for generating civic ideals that would cultivate notions of urban citizenship, a concept vaguely interwoven with ideals of patriotism and a sense of the civic.[10]

During Rudolph Giuliani's term as mayor, the question of urban citizenship acquired a heightened intensity across shifting cultures of patriotism. The issue became of paramount importance as rhetoric issued from the mayor's office triggered disputes over shared civic ideals. Exploiting the discourse of civility, the Giuliani administration grappled with highly contentious visions of civic "harmony." Tension between the heterogeneity of New York's communities and their competing notions of assimilation, set alongside the impoverished ideal of responsible government, delineated the fissures of urban belonging. New regulations regarding such minor infractions as jaywalking, the privatization of public spaces, such as the gardens of the Lower East Side, the banning of street performances, the curtailing of vendors, the reactivation of antiquated cabaret laws, discord with the cabbies, and the corporatization of public spaces cumulatively transformed New York City from a place of possibility to a place of diminished public imaginings.[11] As real estate speculations skyrocketed, the psychogeography of Manhattan shifted. The 1990s in New York City was a time of volatile debates between angry constituencies such as the lesbian, gay, bisexual, transgender community, street vendors, taxi drivers, clubs, street buskers, and the mayor's office.

The Gandhi Garden emerged by the end of the century as one staging ground for multiple and fragile improvisations of urban embodiment. Gandhi stands poised in mid-motion, set apart from the principal narrative of Union Square Park, that of founding fathers and national heroes: George Washington, Abraham Lincoln, and the Marquis de Lafayette. Situated in a park that, since the mid-nineteenth century, has symbolized freedom of speech and the struggle for independence, the location of the Gandhi Garden serves as a reminder of the role of public art in the invention of civic identities. It draws attention to the relationship between myths of nation-formation and international expressions of urban citizenship.

Encountering Gandhi in this particularly frenzied corner of the city elicits a subtle sense of surprise and calm. Stooped, gaunt, but swift in attitude, the modest form is overwhelmed by the garish big box stores that devour the periphery of the park. Gandhi arrests the gaze amid Forever 21, Staples, and a rotating roster of commercial enterprises that have come and gone along with the recession: Barnes and Noble, Virgin Records, Whole Foods, Best Buy. Solitary and resolute, Gandhi offers a thin thread of anticonsumption in a circuit dense with overconsumption.

Frugality in the form of Gandhi confronts the pedestrian drifting through crowd, traffic, and hardscape. His demeanor of quiet determination stands as an anachronism in this age of hypervisibility and narcissistic swagger. Many a strolling person has missed the Gandhi Garden, or only dimly recollects its presence. Yet others stop to chat about the May 2009 auction of Gandhi's frugal belongings: his worn leather sandals, his pocket watch, a brass bowl and plate, and his round-rimmed spectacles, which together sold for $1.8 million.

Viewing Gandhi in silhouette from either side of Fourteenth Street, the observant passerby is struck by the figure's gauntness. Gandhi is a vivid reminder of other regimes of embodiment in the city—self-restrained but alive, self-disciplined yet excessive. We recall Gandhi the long-distance walker, the archetypal urbanist whose personal cosmopolitanism led to a self-imposed lifestyle of frugality and abstinence, experimentation and innovation. Gandhi the advocate and peacemaker. Gandhi the ascetic. Gandhi the organic farmer. Gandhi the yogi. Gandhi the lawyer. Gandhi the revolutionary. Gandhi the nonviolent radical. Gandhi the anti-imperialist, advocate for sovereignty and self-reliance. Most powerfully of all, we recall Gandhi the satyagrahi, the proponent of civil disobedience as strategy for change. All aspects are central to Union Square Park's history of socialist utopian organizing. The Gandhi Statue reinscribes Union Square Park as the seat of an alternative urban imagination within the dreamscape of New York City.

An enlargement in 2003 of Union Square's southwest corner has segregated Gandhi's presence even further. While echoes of Thomas Jefferson, Abraham Lincoln, and George Washington—all sources of Mohandas Gandhi's teachings—bind the Gandhi statue to the themes of liberty, freedom, and democracy embodied by Union Square's public monuments to Lincoln, Lafayette, and Washington, the renovations have succeeded in isolating the solitary figure. Previously caught between two roads, the statue is now no longer visible from any other part of the park. Instead, Gandhi has been given an entirely new, semi-intimate but cordoned-off space that is difficult to maneuver around. The design appears to have crowd control in mind, preventing the possibility of more than a handful of people filing around the Garden. On cold winter days, Gandhi appears to be under house arrest, hemmed in by metal railings, unavailable to the community at large.

Despite obvious efforts to sequester the Gandhi statue's power to activate political sentiments around global issues, the Gandhi Garden continues to be used strategically by various New York communities. Antinuclear activists, ecological urbanists, green market vendors, microfarm adherents, artists, musicians, protestors, antiwar activists, and groups staging civil disobedience are some of the devotees who garland and tend the Gandhi Garden amid the nerve-wracking urban bustle.

Satyagraha as New York Style

Gandhian ideals have repeatedly found resonance, both as political inspiration and aesthetic practice, in New York's post-9/11 climate. From Reverend Al Sharpton to the composer Philip Glass, Gandhi's New York adherents are many. Two instances, a protest and a theater performance, are representative of the myriad ways Gandhian ideals impact New York urban life.

New Yorkers participated in a month of public engagement with Gandhi in April of 2008, as artistic communities promoted the idea of a Satyagraha campaign to perpetuate a metropolitan stance against war and a commitment to sustainability, congealed around Gandhian practices of civil disobedience and nonviolent protests. At the Metropolitan Opera, Philip Glass's opera *Satyagraha* was staged, with billboards advertising *Satyagraha* at major thoroughfares. Gathering around the Gandhi Statue, in Union Square, on April 6, 2008, varying groups of artists, activists, and ecologists, headed by Philip Glass, staged an alternative cry for rethinking the planet's futurity on ethical grounds.

On May 7, 2008, Al Sharpton called a citywide action to protest the acquittal of three policemen who shot an unarmed young black man, Sean Bell, fifty times during his bachelor party outside a Brooklyn strip club, in 2007. Acts of civil disobedience orchestrated by Sharpton at six different junctions of the city's transportation system blocked the flow of traffic. Angry demonstrators of all races converged upon the Brooklyn Bridge, the Manhattan Bridge, and One Police Plaza to protest the verdict. The Triboro Bridge was also barricaded as a security measure.

Sharpton's move to physically impact the city's infrastructure through civil rights strategies of sit-downs, human chains, and the physical mobilization of nonviolent action was a concrete staging of the city's biopolitic in the face of a metropolitan anonymity. Sharpton called the acts of civil disobedience a "slowdown."

Sit-downs, the blocking of roads, and silent protests emphasize that people still matter in the metropolis in ways that one forgets in the minutiae of dailiness. Protesting crowds blocking the Brooklyn Bridge, led by Sharpton and Sean Bell's wife, Nicole Paultre Bell, drew attention to the increasingly polarized sense of urban belonging shaping New York's physical environment, where many black men, in particular, feel deeply unsafe and vulnerable as open, moving targets. A number of men wearing black T-shirts emblazoned "I am Sean Bell" dramatized this sentiment.[12]

In December 2009, Mayor Bloomberg inaugurated a new street name,

Sean Bell Way, amid much criticism and considerable relief across New York City's communities. The symbolic naming of the street where Sean Bell was brutally shot represented a much-needed recognition of the terrible police excesses enacted over the preceding few years. For many supporters of the Sean Bell family, this lip service did not adequately address the continuing imbalance between those who were being frisked and accosted by the police, and those who were not. For the moment, however, the strategies of Reverend Sharpton's civil disobedience and nonviolent methods achieved a certain public recognition of the abuse of coercive police intimidation and force captured in the tragedy of the Sean Bell incident.[13]

Falun Gong Acts

On Forty-Second Street and the West Side Highway, a group of about ten people stood silently, hands clasped in prayer, on a drizzly cold November in 2004. They remained in deep meditation for over an hour, right next to the Chinese Consulate in Manhattan.

In 2005, another group of seven Falun Gong practitioners sat still in lotus positions, in front of Isamu Noguchi's levitating steel *Red Cube*, which was installed in 1968. Situated in front of the Brothers Harriman building on 140 Broadway, between Liberty and Cedar Streets, near Ground Zero, the gravity-defying public sculpture, with its bulky mass poised in flight, delicately balancing its volume against the dizzying heights of surrounding buildings, forms a dramatic backdrop for statuesque activism. Hands clasped in meditation, the group remained in position for at least two hours without moving.

At the tip of Manhattan, in Bowling Green Park, in 2006, facing Arturo Di Modica's famed *Charging Bull*, symbolizing Wall Street's energy, three Falun Gong members in their sixties held posters and distributed leaflets to passersby. The visual contrast between the raging copper bovine and the silent, aging protestors captured quintessential Wall Street contradictions. Rapacious speculative aggression met patient, immigrant will to make the public aware of matters affecting populations in other nations outside the sphere of financial voraciousness.

Facing Ground Zero on a cold Spring Day in 2007, a group of Falun Gong adherents enacted a performance much seen about New York, involving a woman strapped in a cage, being tortured. The blood was red. She was pregnant and hanging. The sight was painful and strange, and no one stopped for too long.

The Falun Gong patiently wait and wander around the public spaces of New York. Sometimes they get into conversations and show an interested passerby how to create a movement. The performers are always obliging when I try to talk to them, but most don't converse in English beyond the simplest of communication, with an emphasis on the word "torture." These exchanges have the disconcerting effect of creating confusion and boredom: confusion about the implications of the performance and boredom at the lack of imagination of the protestors, beyond the same act reproduced at different sites across the city: Times Square, Wall Street, Union Square.

New York is a tough city. Any street performer is competing with bizarre spectacles, such as an individual encased in ice or submerged in a sphere of water, or some other extreme performance unexpectedly unfolding around the corner. Hence, pain has to be entertaining if anyone is going to care on his way to work, or on her return from a long commute. A pregnant woman hanging and being tortured is too obvious, too tacky for New York's taste. Still, it worked. People stopped and stared. They took a pamphlet. They wanted to know more before they picked up their pace.

The Falun Gong is a striking demonstration of a certain manifestation of nomadic urbanism. They transport to the landscape of New York City the memories and experiences of other geographical locales through theatrical reenactments of events whose veracity remain tenuous, scantily substantiated episodes of torture intended as propaganda against mainland China.

Mass enactments of physical gestures, set against the backdrop of New York City, give the Falun Gong a New York veneer. They come across as displaced protestors with a cause that no one understands, appealing to a population that has no interest in the immediate future of the Falun Gong in China, or in New York City. Yet they also fit into a genre of protest and activist performers whose causes are little known, but whose physical presence within the city is part of the urban landscape. The demonstrators who position themselves randomly at key traffic junctions, with placards against the Iraq war, are one such spontaneous manifestation of activist actions in the city that are enacted with the sense that their performance is for the world city as a stage.

The assumption underlining these urban performances is that New York City is a center of power. However, there is no chosen space in New York where power can be influenced. Consequently, for many activists, crowds equal influence and political impact. In reality, most of the time, these public displays are momentarily thought-provoking without causing deeper repercussions.

The urban enactments of the Falun Gong fit into a New York tradition that Jane Jacobs identifies as "unslumming." Unslumming, writes Jacobs, "begins with those who make modest gains, and with those to whom personal attachments overshadow their individual achievement."[14] Jacobs is referring to the process of human interactivity that dynamizes the beleaguered neighborhoods referred to as slums in the 1950s. However, this process of unslumming is also a form of engagement that interrupts neighborhoods that are diverse and bustling with commercial activity.

Unslumming, as Jacobs defines it, is an urban phenomenon of rearranging social patterns within an urban setting. It can happen within the scope of a block or a building. The Falun Gong's activities demonstrate how unslumming can be a diversifying social activity outside the locus of a disempowered neighborhood. As an immigrant population, the Falun Gong straddle multiple communities within New York City. They interrupt assumptions about the urban underclass, which includes large numbers of Chinese immigrants, as they find time for public activism. To that extent, they break the stereotypic impression of Chinese immigrants' being part of an indistinctive Chinatown. Their nomadic presence around New York City, as an activist group, makes their immigrant intervention a dynamic and active aspect of the urban landscape. The Falun Gong bring a liveliness to the streetscape that is temporarily animating.

For Jacobs, the city of New York invites any number of improvisations of the self, and of the received interpretations of cultural identities. The case of the Falun Gong highlights the tensions between national identities, such as that of being Chinese or American, and the complex negotiations between migrant and immigrant identities, such as those of the Falun Gong diaspora in New York and elsewhere across the world. The Falun Gong are seen within China as insurgent terrorists, and as dissident and insurgent against the Chinese American community in the United States. In New York, their regional American identity foregrounds the intricacies of how immigrants, refugees, and political asylum-seekers, as many Falun Gong practitioners claim they are, impact the urban fabric of the city. The Falun Gong combine Jane Jacobs's notion of unslumming and a literal merging of their lives with their causes, making a lifestyle of it on the streets of New York. In the process, they make a spectacle of a personal, indefinite metamorphosis.

ecological expressivity

Ecological urbanism, the author David Owen points out in *Green Metropolis*, is an idea filled with conflicting ideologies, commonsensical assumptions and self-serving marketing ploys. Owen provocatively lays out the numerous ways the very idea of sustainability is fraught with widely divergent understandings of what environmentalism in an urban setting entails.

Contrary to the pedestrian assumption that living in Vermont is "greener" than dwelling in a metropolis, Owen persuasively argues that the simplest principles for a more energy efficient lifestyle are to live closer, live smaller, and drive less.[1] Owen demonstrates the inherently unsustainable energy implications for living far apart, with the expectation to drive further, entailed in suburban and rural lifestyles. Owen furthers his iconoclastic view by closely comparing the economies of scale involved in energy consumption in cities and suburban contexts.

"Ecological Expressivity" applies David Owen's approach to expanding the idea of sustainable living to include the role of social performativity in creating one of the quintessential elements of political cosmopolitanism, which, according to the political philosopher David Held, is the ethical imperative of sustainability.[2] Held identifies the principle of sustainability as a fundamental concept of cosmopolitan citizenship, drawing upon Kant's precept of cosmopolitan law, which states: "This right to present themselves to society belongs to all mankind in virtue of our common right of possession on the surface of the earth on which, as it is a globe, we cannot be infinitely scattered, and must in the end reconcile ourselves to existence side by side."[3] According to Held, all economic and social development must attend to the protection and preservation of the world's core resources. Held demarcates these resources as those that are "irreplaceable and non-substitutable."[4]

This section takes Held's position and suggests that cities are "non-substitutable" living environments. They become zones of core resources with finite living spaces and limited access to basic amenities, such as water, air, energy, space, sanitation, and waste management. Human bio-power is the critical resource from which workable ideas of sustainable growth must evolve. The role of social performativity is to cultivate the cosmopolitan principles that Held lays out as active agency, the creation of democratic consent, and the inculcation of responsibility.[5]

One mechanism for animating such highly contentious communicative processes is the role of performative engagements within public space. Public performative acts invite debate and dialogue. As the political philosopher Martha Nussbaum suggests, instead of contemplating impending ecological horrors, these interactions become opportunities "to act, doing something useful for the common good."[6] They provide conditionally noncoercive provisional spaces for encouraging active agency.

The following section addresses instances of performative engagements that become venues for public opinions and discursive visions of how a city's finite resources of space and energy can be mobilized toward durable ends. Urban gestures in public spaces invite processes of accountability. They become conduits for the democratic display of personal responsibility, through which a new kind of sustainable city is being reimagined in New York.[7] This transforming city is one that must increasingly be hospitable to widely divergent demographics, searching for forms of participatory dialogue that allow for different political communities to coexist within an island-city with finite space.

CHAPTER 7

greening hardscape

In 2013 Manhattan's waterfront is an expanding, green space of public parks, human pathways, bike trails, and piers reclaimed for neighborhood use. Children hang from monkey bars, people stroll, skate, bike, run, kiss, sunbathe, stare, cry, brood, and meditate by the splendid new stretch of green spaces stretching from the tip of the island all the way up the west side of Manhattan. The city is recovering its waterfront in ways that were never imagined in the previous era of heavy shipping and maritime industry, an era that has been on the decline since the mid-twentieth century. Former industrial waterfronts fallen into neglect and disrepair now contain mixed-use neighborhoods combining residential with commercial and park amenities. The transformation of the city's shoreline from shipping-related activities to social gathering places serving residential neighborhoods has redefined New York's waterfront as a pleasurable destination for its communities.

The momentum for the greening of New York City, built under Mayor Bloomberg's administration, is facing great challenges, however. Budget cuts in 2009 have affected the refurbishing of school playgrounds into projected community green spaces, along with other earmarked park expansions, setting back the city's projected goal to expand access to leisure spaces for all communities in the city.

Faced with New York's fiscally strapped coffers, the city would do well to remember an earlier era in the greening of New York City to revive its economically devastated imagination, that of New York's admirable culture of community gardens and public gardens. Our volatile history of hard-fought efforts to create and retain green spaces within the city, at a time when there was little green space to alleviate the stress of urban living, offers many lessons in the creative reclamation of green spaces.

Two symbolic gardens from this earlier period of green activism serve as instrumental reminders of the necessity for park space in urban life. The dramatic and tragic demolition, in 1986, of Adam Purple's famous public garden, called the Garden of Eve, located on Forsythe Street, and later the violent and brutal demolishment of the famed and culturally iconic garden called Chico Mendez, in 1997, exploded the public's awareness of the importance of community gardens to a neighborhood's well-being. They paved the way for a heated public debate on the need for and nature of green spaces in New York City, and in Manhattan in particular.

Public Space and the Uncivic City

It was a hot July day in 1998 in New York City. A group of concerned citizens and Lower East Side activists, a hundred strong, gathered on the steps of City Hall to protest Mayor Giuliani's corporate agenda of privatization and gentrification. More specifically, the protestors had come to denounce the sale on July 20, 1998, of four community gardens and a city-owned building that had been rehabilitated and occupied by Charas/El Bohio, a Lower East Side community organization. The sale was ordered by the Giuliani administration, along with hundreds of other "vacant" lots and buildings citywide.

This real estate assault on the poor and working citizens of New York—dramatically illustrated in legislation against sidewalk vendors, cabbies, bicycle messengers, community gardeners, and street performers—had become a routine feature of life in New York during Giuliani's "quality of life" mayorship, between 1992 and 2001. Ordinances criminalizing urban pleasures, jaywalking, performing in the streets and subways, selling food on the sidewalks, accompanied by the insincere rhetoric of civility used to present various gentrification and privatization measures, generated a sense of uncivicness that was troubling to a city that has always prided itself on its local myths of urban awareness, neighborhood, and community. "Even the rich cannot live in New York anymore," proclaimed one protestor at City Hall.

Can feelings have a history? This was the question asked by the historian Bronislaw Geremek in his study of medieval poverty.[1] Certainly the vocal and creative protests on July 20, 1998, by artists, squatters, community gardeners, activists, and concerned citizens opposed to the sale of community centers, community gardens, and public lands suggest that the feelings attached to particular buildings, lots, and public spaces extend beyond the geographical pertinence of a specific site to its neighborhood. The emotions generated by the sale of Charas/El Bohio, described variously as a place that since 1979 had become crucial to the Lower East Side's Latino community, indispensable to the artistic community, and important as a base for many underprivileged members of the neighborhood, bring Geremek's question to the fore as an urban structuring idea.

The relationship of Charas to the feelings associated with the Lower East Side, and the city more generally, encapsulates two decades of public reclamation work initiated by local residents in the revitalization of neglected neighborhoods and abandoned buildings. According to Vivian Miller, of the New York City Community Garden Coalition, many of these sites were discarded by the city in the 1970s and left to decay. The abandoned sites became garbage dumps, accumulating toxic waste and rodents and attracting drug traffickers to the area. In response, gardeners living in the community volunteered to become stewards of the neighborhood by reclaiming these lots as green spaces to be used for public enjoyment.[2]

This reclamation process extended to abandoned public buildings. Charas appropriated P.S. 64, a former public school situated on Ninth Street, east of Tompkins Square, in the interest of urban artists and community recreation. The organization began squatting the building when it was an enclave for sex workers, drug dealers, and street people. Hence, Charas's social impact on the neighborhood and its importance as an enactment of a civic ideal extended far beyond its local pertinence as a cultural and performance center. The centrality of Charas as a cultural and local landmark operated at the level of an institution within the city. It offered some of the most affordable rehearsal space for hundreds of artists and cultural workers, and provided a physical base for the rehabilitation of the heterogeneous Latino neighborhood through culture and entertainment. Charas also served as a youth center and offered after-school programs, along with gallery, film production, and exhibition spaces.

The protestors' challenge to the city, led by Armando Perez, Charas's artistic director and the East Village Democratic district leader, hinged on the importance of a public property to a particular locality prior to sale. The question raised by Perez's group is crucial to understanding how the feelings ascribed to a particular site can accrue a sense of historicity. In the case of Charas, this history of feelings resounded at the intersection of urban civic commitments, citizens' rights to the city, and the city's responsibility as an arbiter of civic authority when negotiating public lands and private ownership.[3]

The array of issues raised by concerned community members leading up to the sale of Charas brought together a network of civic commitments involving artists as citizens and the meaning of public property in an urban context. The first issue, as pointed out by Perez's organization, was the function of a site in relation to the changing demographics of its neighborhood. Another issue was the investment of labor, time, and money—often made at great sacrifice by community members—necessary to keep the organization afloat. The city and the buyer, the developer Gregg Singer (whose name the Giuliani administration refused to disclose), were capitalizing on this public investment for private profit. Finally, the group contested the high-handed manner in which the sale was made, with no input from local residents, and called for a moratorium on the sale of public land until a democratic process incorporating all involved parties could be agreed upon. Not surprisingly, this demand, which was supported by a variety of local constituencies, such as advocates of low-income housing, community organizations, gardeners, cultural workers, and concerned residents, was ignored by the city.

The sale of Charas and the subsequent tragic death of Armando Perez, its charismatic director, reiterate the city's increasing public disavowal of low-income, immigrant communities and the multiethnic communities of the Lower East Side, and of New York in general.[4] The rhetorical effort to appeal to the city's magnanimity, through the careful use of civic language, such as "good faith negotiations," with the implication that such an arrangement is beneficial to the city as a whole, ultimately proved futile. Notions such as "vital cultural and community center" and "good faith" are anachronistic operating logics in the face of corporate capital. Charas/El Bohio's dismantling symbolized the clash of community sentiment and corporate greed that taints the city's exhilarating skyline. Manhattan's corporate avarice mars its myth as a city where anything is possible.

Gardens versus Community Housing

According to Todd Edelman, of the Garden Coalition, who actively protested the sale of Charas and four other community gardens in 1998, various constituencies articulated the struggle over public land and privatization as a fight between community gardeners and housing advocates who were competing for scarce goods. This misinformation circulated in the interest of dividing the two groups and weakening their shared sense of concern, noted Edelman.

The properties advertised as "vacant" lots were in many cases plots that had been abandoned by an indifferent city since the 1970s, then reclaimed by residents and gardeners for public purposes stemming from neighborhood needs. Many of these spaces have since become gathering places for immigrant, ethnically diverse, and minority communities in the city. The enjoyment and benefits derived from these green retreats cannot be quantified. Just the sudden pleasure of encountering a green oasis of carefully nurtured plants and flowers in a relentless hardscape is restorative to the spirit.

At a deeper level, a "history of feelings" is encapsulated in a community garden, including the civic commitment to one's neighborhood and city, the specific social relations inscribed on specific locales, and the religious and mystical sentiments attached to working and revitalizing the earth. Cults of earth goddesses, such as Gaia, the Virgin of Guadalupe, and Persephone, inform local New York garden subcultures as much as the civic-minded challenge to corporatizing private lands.

Despite such community investment, city agencies embarked on a policy of selling off reclaimed spaces to the highest bidder.[5] This move to privatize public land was speeded up by the termination of "green thumb licenses" by the Department of Housing, Preservation, and Development. Green-thumb licenses, generally issued for periods of eight-and-a-half years, were held by seven gardens in the Bronx, eighteen in Brooklyn, and thirty-two in Harlem. With the termination of these licenses, in 1998, thirty-two gardens on Manhattan's Lower East Side came under the immediate threat of being sold to developers of luxury condominiums and commercial enterprises. The implications were far reaching. Not only would this new incentive usurp public space, build over what had become green retreats, and devour much-needed green space in the city, it would also reduce the range of social services that community gardens offered free of cost as civic spaces in the city. Privatization initiatives left green spaces on church grounds as the last remaining enclaves of green refuge.

Ultimately, the inevitable gentrification would displace already peripheralized low- and middle-income populations. In July 1998, the city sold parcels of land with gardens to buyers at phenomenally high prices. The city claimed that the parcels would be used for low-income housing. Buyers flatly contradicted the city's claims by stating lots would be developed expressly for Silicon Alley, as the dot-com bubble was known, and Wall Street markets. In the July sale, the Puerto Rican community lost three important gathering places: the Villa Vitin Garden, on East Fifth and Avenue D; the Fourth Street Casita Garden, on East Fourth, between Avenues C and D; and the Lower East Side Hispanic Community, Inc., Garden on East Fifth and Avenue D.[6] As described above, this drastic reduction of green retreats was spun in the political arena as a turf war between community gardeners and low-income housing advocates.

The years following the struggles to save the community gardens of New York City have witnessed a massive explosion in high-rise construction on formerly "abandoned" lots, interrupted by a massive building halt. Now these lots are not only neglected, but dangerous, as half-constructed buildings disintegrate into disrepair. The nationwide economic depression has strangled New York at all levels of the economy, with its dramatic loss of jobs and rising unemployment. Amid this grim scenario of a recession without an end is a growing public awareness of global climate change as a local issue. Within New York City, the conversation has shifted from a literal involvement through the informal activities of neighborhoods engaged with their vacant lots, to a citywide effort to build more parks, plant more trees, and establish more car-free zones, a variety of green initiatives involving earth, water, and air.

Air toxicity, beginning with the Clean Air Act of 1990 and the subsequent efforts to promote clean air through the use of clean fuels, opened up the issues of environmental well-being and the ecological impact of metropolitan regions such as New York City. Through the concerted promotion of sustainable architecture and design, New York has slowly initiated a more ecologically conscious approach to urban growth. The advocacy of adaptive reuse for the purposes of generating a greener city is perhaps one of the city's more visible efforts at impacting densely developed areas, particularly downtown Manhattan.

Learning from the practice of urban gardening techniques explored by the Lower East Side gardens, many new communities have picked up the mantle of sustainable land use. The expansion of rooftops as urban roof gardens, the cultivation of small ecosystems, such as rain gardens, the microgardens on rooftops around Union Square, integrated water systems, the adaptive reuse of an abandoned train line for a public park, the adoption of new materials and moving away from the use of Brazilian teak in bridge renovations in New York City are some forms of green urbanism impacting the city.

Robert Moses pursued the particular problems facing a narrow island of rising global prominence as opportunities to create the most interesting built environment on the planet. For Moses, Colonel Egbert Viele's map represented an invitation to expand, rather than a reminder to reconsider the delicate ecology of this unusual island city. During his stint as commissioner of parks, Moses went on to build over three hundred children's parks all over New York City, creating the first generation of public play spaces for people across New York City.

Along with his public playground project came Moses's large-scale public project to establish leisure spaces across New York City's waterfront, through the creation of large beaches approachable by his highways, such as Jacob Riis Park, Far Rockaway Beach, and, with some difficulty, Jones Beach. Yet, at the heart of Moses's projects was a rejection of the promise of New York City's waterfront, demonstrated by the locations he chose for the Brooklyn-Queens Expressway and the West Side Highway. These two large highway projects effectively cut off the people of Manhattan and Brooklyn from using their waterfront as a resource for relaxation. Instead, he turned some of the most spectacular vistas of waterfront into a passing spectacle for harried commuters. The waterfront effectively emerged as dead space, dangerous to access because of heavy traffic. Piers around Manhattan were abandoned as industrial shipping was discontinued. The waterfront became desolate through lack of mixed-use habitation.

Under Mayor Michael Bloomberg's PlaNYC 2030, the city is consciously transforming its infrastructure at all levels—land, air, water, energy, and transportation. The manifesto proposes a vision for ecological sustainability and the reclamation of New York's waterfront as a space for enjoyment and public leisure. Bloomberg's public message to reduce the carbon footprint of technologies and commodities in the city has encouraged a preliminary conversation on the very controversial topic of congestion pricing. Congestion pricing was a proposal to levy tolls on vehicles driving into Manhattan from New Jersey and Brooklyn during peak hours. Considered elitist and antibusiness by some, the proposal was favorably received at the city level and knocked down at the state level, with great pressure from the State of New Jersey, whose residents compose a large percentage of the commuters to New York City.

Between 2006 and 2013, Bloomberg's PlaNYC 2030 and its greening initiatives, such as MillionTreesNYC and the Summer Streets project, which opened up New York as a car-free environment for three weekends in August, for five summers in a row, 2008 to 2012, woke up New Yorkers to the possibilities of a pedestrian city. The MillionTreesNYC project alone planted over 220,000 trees in New York City, a much-needed effort to bring more green to more neighborhoods.

No More Fake Parks

Following the promise and optimism of PlaNYC 2030's manifesto in 2007, a downturn has affected Bloomberg's ambitious plan to reduce New York City's global warming emissions by 30 percent by the year 2030. Bloomberg's announcement of massive budget cuts at the end of 2009, following global financial collapse, has interrupted the multiple strategies of greening under way, including transforming school playgrounds into park space, creating green and white roofs, expanding New Yorkers' access to their waterfront, and forcing New York to adopt more energy-efficient buildings. One of the sad, but perhaps revealing, effects of the terrible budget deficit of New York's economy is Mayor Bloomberg's wavering commitment to his environmental promises.

One of the first items to suffer cutbacks was the parks initiative, a project to convert existing school playgrounds in the Bronx, Queens, and Brooklyn into much-needed neighborhood park space that remained open after school hours. Many people were never convinced that this proposal represented a serious effort to effect change. It was viewed more as a half-hearted attempt to infuse neglected neighborhoods with fake turf to pay lip service to environmentalism. In the few parks that had "renovations," the work was so poorly executed that the new "greening initiatives" did little to improve the quality of life for neighborhoods without access to park space.

The issue that most decisively set back Bloomberg's contribution was the problem of reducing carbon emissions in the city. In 2009, Bloomberg backpedalled on legislation that would have forced buildings of over 50,000 square feet to become more energy efficient. This legislation would have forced existing buildings and future constructions to be audited and would have required them to adhere to energy-efficient standards. Known as the Greener Greater Buildings Plan (GGBP), the legislation has now been reduced to an optional suggestion, with compliance left to individual developers and landlords. Buildings contribute to 80 percent of New York City's carbon emissions, and the GGBP legislation would have forced a historic shift toward a more environmentally committed city, as well as provide a model for other American cities.

This is a tough time to persevere in expensive legislation that makes carbon emission control a mandated undertaking. Nonetheless, New York City has learned that it has everything to lose by deflating momentum on the greening initiatives of the densest urban conglomerate in America. New York will continue to pursue the battle for GGBP legislation, even if it has to compromise on some of its key reforms. Creating energy efficient buildings is not a choice, but a necessity for the future of dense living.

Gardens in Transit: Cabs, Train Tracks, and Piers

Purple, green, and pink flowers, florid magentas and brilliant blues, sped by during the fall of 2007. The yellow cab had become an advocate for New York's renewed commitment to the greening of the city, heralded by Mayor Michael Bloomberg's PlaNYC 2030 goal that "all New Yorkers should live within a ten-minute walk of a park." Vibrant-colored gardens decorated the roofs and surfaces of New York City cabs.[7] This declaration of green love for the city's public spaces was a strategic wakeup call to urban dwellers that there is not adequate green space for every inhabitant in this city of eight million.

The year 2007 was the 100th anniversary of the motorized and metered New York City taxi. As part of the centenary celebrations, the city launched a massive public art project called the Human Flower Project, a brainchild of two brothers from California, Ed and Bernie Massey, who run a therapy organization called Portraits of Hope.[8] Begun in 2006, the project involved 23,000 public school children who painted plastic decals over a period of a year. These decals were designed to be pasted onto the exteriors of taxis to commemorate the centenary celebration of the New York City Taxicab. The effect of "flower power" on the streets was tangible: people were elated, curious, horrified, bemused. The project's goal was to generate public awareness through the city's least protected citizens, the children of public schools, and create a hopeful and environmentally conscious city.

The Gardens in Transit project signaled a shift in direction for New York City's development. The greening process had been steadily gaining public attention until the 2009 recession, reminiscent of the fiscal downswing of the 1970s. The rise of New York City's public gardens from neglected, vacant lots to thriving, urban green spaces paved the way for imagining the city as a potentially expansive green space that hasn't fully been realized.

Taking its cue from Gardens in Transit, on April 22, 2008, the Metropolitan Transit Authority of New York City issued five million limited-edition metro cards with leafy fonts and eco-friendly sound bites. The cards were not made from recyclable materials. However, their green symbolism indicated a growing, though insufficient, commitment on the part of the Metropolitan Transit Authority to increase its use of renewable resources, including solar and wind energy. Following the plastic flower decals and leafy metro cards, neither of which were biodegradable or energy efficient, the Metropolitan Transit Authority's Blue Ribbon Commission on Sustainability announced in 2008 that by 2015 it planned to generate 7 percent of its energy through energy-efficient technologies, such as solar panels and wind turbines. The harnessing of tidal energy from turbines in the East River is projected to generate a considerable portion of the power needed for the F train's Roosevelt Island subway station.[9]

Transportation such as taxis, buses, and other transit systems are major contributors to pollution. New York's slow attempts to reduce its congestion problem by incorporating hybrid cabs and buses and expanding bike access and public interest in biking have been much-needed inroads. Public awareness about reducing the carbon footprint of transportation networks has increased demands for transportation alternatives. The call for the Metropolitan Transportation Authority to expand bike parking inside subways in New York City, along with the now dissolved plan to reduce the flow of traffic in downtown Manhattan through congestion pricing, a system of tolls for cars coming into the city, are some of the recent debates that address this new awareness.[10]

The idea of greening New York City through the literal greening of concrete and macadam with tree plantings, and the technological greening of the city's transportation networks, has, to date, remained a rhetorical move that needs a lot more infrastructural support from the city and the state. The sentiments of many involved in creating alternative lifestyles geared toward reducing the city's carbon footprint, such as the Lower East Side Ecology Center's composting program and the Transportation Alternatives advocacy group, agree that the city needs to do much more to achieve Bloomberg's projected goal for reducing congestion and carbon emissions in the next few years. Merely using energy-saving light bulbs in municipal and public buildings, such as Grand Central Station, displaying eco-friendly advertising on public trashcans, providing composting facilities at the local farmer's market in Union Square, and reducing the sale of plastic goods across the city will not produce the cultural shift away from hyperconsumption and toward a more sustainable management of the amenities and needs of urban dwellers.[11]

Between Water and Green Space: Viele's Dilemma

Aerial views of Manhattan dramatically showcase the jewel in the built environment of New York City: the expansive, precisely planned lungs of the city, Central Park. One sees a vast, open area of carefully planned green space framed by the water around the concrete island that defines Manhattan's identity.

The Commissioners' Plan of 1811 had no provision for such a large park. The emergence of a park on such a massive scale, on so tight an island, emphasizes the centrality of the city's unfolding identity as an artificial environment. The need for a large park arose by 1853 from the real problem of population density and its attendant pestilence of yellow fever, which gripped New York intermittently as a result of crowding and little green space. Described by Dutch settlers of the seventeenth century as a verdant, green island, Manhattan had diminished its green spaces considerably by 1865, ceding them to an expanding concrete jungle built over streams, rivulets, brooks, and swampy ponds. This rich ecology of water and forest that structured the city's topography before Dutch colonialism had dramatically shrunk to patches of green in Colonel Egbert L. Viele's map of 1865.

Colonel Viele, the engineer who designed the sanitation system for Central Park, reconstructed what is known to be the most detailed articulation of hydrology sources and green spaces in the Manhattan of the nineteenth century, painstakingly demarcated through then-available records, in order to precisely situate Central Park's dimensions within the geography of Manhattan. Viele's remarkable map is a cartographic rendering of what has continued to be the single most challenging endeavor facing the island of Manhattan: sustainable development in an ecologically fragile environment with finite land space.

Viele's map of 1865 raises many issues that are being furiously contested today. Viele clearly demarcates water sources buried under built environments, rivers and streams covered over with concrete and stone. His map evokes the question for our time: How much development can the island take before it becomes inhuman? How might builders and planners draw upon existing environmental compositions to produce a livable city with a smart-energy future?

As the city's population expands to over eight million people with diminishing green space and a fervent rush of overbuilding, Viele's dilemma raises the point of ecological vulnerability once again. The only green and yellow seen in many neighborhoods in 2007 flew by on the tops and sides of cabs in saturated neon colors, mere promises of what the environment can be—and once was, but now built over.

A Park for Every New Yorker

In 2007, PlaNYC 2030 was presented to the city of New York as a manifesto that included a push to green the city and bring a park within a ten-minute walk of every New Yorker. Announcing the plan, Mayor Bloomberg stated: "You cannot think about transportation without thinking about air quality. You cannot think about air quality without thinking about energy. You can't think about energy, or any of this, without thinking about global warming."[12] Bloomberg's call to revitalize the city through energy-efficient restructuring has redefined the city's identity. There is a decided shift away from its established persona as a city of concrete toward a more contested sense of a city's rethinking its carbon emissions. Slowly but surely, the public impetus to bring ecological sustainability into the discourse on development and the redesigning of urban spaces has created a new awareness around how people occupy space in New York.

The trend toward a people-centered approach to planning emerges from the techniques established by Jane Jacobs in the 1950s, in her condemnation of top-down city planning. Using Jacobs's model of mixed-use neighborhoods and low-intensity development, a model geared toward how people on the street use space, many conversations since the 1970s in the city have contended with the tension of large-scale urban planning, implemented by Robert Moses, who held a grand vision for the city, versus local, proactive engagement, which calls on residents to identify their needs in relation to the plans being draw up for development.

A plan in the 1970s to move the elevated Westside Highway underground and turn the waterfront into parkways, called the Westway Project, became a major topic for contentious debates between large-scale development initiatives and neighborhood concerns prioritizing use and environmental sustainability. The Westway Project was fought vociferously by neighborhood activists and environmentalists at the time. One of the many objections to the top-down planning measure was that it bore the heavy hand of a Robert Moses-style approach to development. Citizens and activists concerned with preserving the identity of historic neighborhoods wished to prevent the kind of massive gentrification that has now completely gripped the west side of downtown Manhattan.[13] In the end, it was the environmental impact on the Hudson River's striped bass population that prevented the Westway from being built.[14]

The shift in balance between local activist voices, commercial entrepreneurs, and the rapacious intentions of investors whose concerns are mercantile, rather than civic and public-minded, has played out along the West Village's waterfront in questionable ways. In 1998, Mayor Giuliani and Governor Pataki initiated plans for the development of the Hudson River Park, a green beltway extending from West Seventy-Second Street to the Battery, in downtown Manhattan. Subsequently, the Lower Manhattan Development Corporation rose out of the ashes of the devastation of 9/11 and its grim aftershock as a player in revitalizing downtown. The debris, ash, architectural scars, and infrastructural damage led to a massive reconfiguration of the waterfront identity of downtown Manhattan, concurrent with the greening of the west side of the island.

The Hudson River Park project is a case in point, where many questions raised by environmentalists, urbanists, and activists, questions about the economic, cultural, and planning implications for climate-change adaptation, have played out. Close interfacing between private donors, city agencies, public interest groups, state and federal monies, local community members, and civic and municipal authorities created the Hudson River Park Trust, a city-state organization that constructed and maintains the park.[15]

The initial successes of the Hudson River Park Trust to mobilize private and public funds concealed hidden concerns highlighted in the subsequent realization of the Hudson River Park. Legislated by lawmakers to be a self-supporting undertaking financed by the profits of Pier 40, a mixed-use sports and parking facility, and Chelsea Piers, a multisport commercial venture, the cost of financing the construction and maintenance of the park has ballooned without proportionate funding sources to pay for its upkeep.[16] Hopes of turning Pier 40 into a profit-generating entity have been stalled as its decaying infrastructure has led to the shutdown of a large number of its revenue-generating parking spots. In addition, constraining restrictions on the types of development Pier 40 can incorporate has stymied its commercial prospects. Consequently, a potentially utopian green development is gripped by the very contradictions embedded in a green metropolitanism: how to make eco-friendly urban planning financially eco-sustainable. As many critics have noted, a mandate to green urban space needs to be accompanied by a plan to activate industry and production around such projects. This challenge is a balancing act that cities like New York have yet to successfully address.

Another case in point is the unsustainable scale of pedestrian activity that has erupted adjacent to the extended parkland for public mixed use, established by the Hudson River Park Trust entity. A new influx of development has emerged in the Far West Village, dramatically altering the historic waterfront's industrial traces, creating high-end consumption, without a substantive growth in industrial production-oriented entrepreneurship. As a result, the West Village and Meatpacking Districts have become glaring markers of the tension between greening incentives and existing industrial areas. They demonstrate how commercial growth is sometimes in tension with the preservation of New York's industrial districts. In this case, the transition of industrial infrastructure to recreational use raises the question: What about the loss of industry implicated by this radical re-

arrangement of space from one of labor to one of leisure?

Around historic Gansevoort Street, which marks the ancient footprint of the Native American settlement that existed around a cove famous in the sixteenth century for its oysters, such a transformation of commercial activity from low-intensity industrial use to high-octane nightlife has created a standoff between commercial investors, neighborhood residents, and older commercial business owners, such as meat packers, intent on maintaining, and now bemoaning the loss of, the character of the area.[17]

Tom Wolfe's nightmare glass and steel box city has come to roost in Manhattan's last gritty industrial neighborhoods in the Far West Village. The unfortunate destruction of the sole remaining factory building in the West Village, belonging to Superior Ink, is a testament to the potential fate facing other low-lying industrial neighborhoods. The Superior Ink building was an ochre-bricked nineteenth-century factory with a brick chimney that the Greenwich Village Society for Historic Preservation and the Landmarks Preservation Commission fought to save. Once demolished, it was replaced by a high-end residential brick behemoth. Superior Ink is a reminder of what will continue to happen at a ferocious pace if the Landmark Preservation Commission's efforts to designate unprotected low-lying industrial neighborhoods with more stringent development criteria do not proceed with urgency.

Mayor Bloomberg's ambitious promise of "a park for every New Yorker" is a much needed one. However, it requires an imaginative exercise of private and public funds, educated human resources, and effective governance structures for the cultivation of green spaces across economically uneven neighborhoods. Cosmetic transformations along wealthy neighborhoods do not address the deeper implications of greening a large metropolis with swaths of city blocks without green spaces in sight. The real hurdles in achieving a more comprehensive eco-efficient change in New York lie in educating less robust neighborhoods in the techniques of community organizing and fundraising for local goals. This task is one that haunts urban communities in First and Third World cities, across geopolitical boundaries.

Current sentiments toward Mayor Bloomberg's investments in greening the city reflect the opinion that what is really under way is merely suburbanization. The public is ambivalent about what is generally viewed as the discrete production of a new kind of elitism, one based on the cosmetic greening of New York's public spaces.

Despite public reservations about the economic impact of the reclamation of waterfront as park space since the collapse of the World Trade Center, there is a distinct shift in the spatial configuration of downtown Manhattan toward a new aesthetic of public enjoyment. It has generated new competition for public spaces such as the Christopher Street Pier, where diverse communities coexist, using the pier for a variety of social purposes. Recent years have witnessed an escalation of conflicts between neighborhood residents and the transient groups of youth who travel from around the New York-New Jersey area to inhabit Christopher Street Pier well into the wee hours of the morning.

Conflicting interests between residential neighborhoods bordering the pier and the multiple communities of youth adopting the pier as a lifestyle destination have raised issues of maintenance, crowd control, surveillance, safety, noise control, economic disparity, and rights to public space. What was once a subcultural destination for New York's gay community has over the years transformed into a highly surveilled, mixed-use space of great popularity. With gentrification came well-heeled lifestyle interests that conflict with the vulnerable populations the pier attracts nightly.

Christopher Street Pier is a case where a waterfront park space is loaded with the pregentrification history of the site's cultural usage as a predominantly gay hangout.[18] Transforming Christopher Street into a livable, multidimensional park space with ample room to accommodate a variety of social lifestyles has reconfigured the assimilation of gay New York as part of the urban fabric of everyday life. Children and the elderly, people in wheelchairs and toddlers, coexist with bands of LGBT groups enjoying the expanse of the city's free spaces.

One spectacular project under way on the west side of downtown Manhattan has dramatically transformed the Manhattan waterfront: the much-debated High Line project, which opened to the public in 2009. Mayor Giuliani wanted to demolish the High Line, a WPA-period elevated railroad track running up Tenth Avenue. Giuliani's successor, Mayor Bloomberg, became a major advocate for the High Line project, which was conceived and developed by a private entity, the Friends of the High Line. The City of New York took ownership of the High Line in 2005.

Consider grass, flowering plants, and native shrubs in the midst of urban density. Imagine a park above transportation gridlocks. Picture open sky between metal pathways and concrete built environments. Visualize lounging people on wooden benches in the crisp, biting sunlight of November. This is the marvelous gift to New York City that is known as the High Line.

The High Line was built in the 1930s for the transit of industrial goods. It was closed down and parts of it were dismantled in 1980. The few sections of the High Line that remain have now been transformed into an urban parkway that cultivates existing flora found on its formerly abandoned train tracks.[19] The High Line is the first elevated public park of its kind in New York City and in the United States. It is hailed as a poster child for the future of adaptive reuse as a development strategy in densely populated urban spaces. Taking the structure and ecology of the abandoned railway line as its aesthetic, the Friends of the High Line undertook the imaginative pursuit of transforming an abandoned urban site into a roof-garden complex that stretches from the west side of downtown Manhattan all the way to Thirty-Fourth Street's Penn Station.

Transforming industrial wastelands into dramatic parkland and accessible waterfront public spaces can generate new urban dilemmas for neighborhoods formerly structured for industrial use. The High Line's success is a case in point. A large influx of crowds streaming through the Meatpacking District and surrounding environs as a result of the demand to experience the High Line has put surrounding low-lying residential neighborhoods under enormous strain. While the famed public park enjoys the attention it is garnering for its outstanding design concept and landscape architecture, the vicinity has been propelled into a new traffic hell, as large tour buses and heavy pedestrian traffic weigh down its cobblestone streets.

The gift of the High Line to the West Village has also exposed a formerly quiet, residential area to a massive reduction in quality of life. The escalation of street noise, vehicle traffic, pedestrian traffic, and the heightened influx of crowds to an area not designed to accommodate such large numbers of people have caused the deterioration of the aspects of the neighborhood most valued by the community: peace, quiet, and neighborhood feel. The counterproductive impact of this particular green initiative raises larger implications regarding infrastructure, traffic congestion, and neighborhood maintenance issues when destination parks are imposed upon formerly low-visibility neighborhoods. A high-concept park without a plan for neighborhood integration can generate considerable local dissatisfaction.

In contrast with high visibility parks such as the High Line, smaller park projects are being developed on the west side of Manhattan as gardens in transit, sites people can pass through while enjoying splendid vistas and moments of contact with the shoreline of Manhattan.

Pier 25 used to be a popular, seedy public park on the waterfront, with a kiddie golf park, an amphitheater, and a sandy, much-used volleyball court. From the boathouse moored on its premises the community could enjoy free kayaking. Watching it being razed to the ground brought melancholy and sadness to many a child and parent who had visited its ramshackle premises.

Pier 25 has been transformed to contain more park space, albeit corporatized in its aesthetic thrust. The pier is designed as a multipurpose park with volleyball courts, a miniature golf course, a café, a children's playground, and an estuarium. Part of the Hudson River Park complex that stretches from Tribeca to Midtown, Pier 25 offers another instance of a decaying pier that has been reclaimed for sport, walking, running, bicycling, picnicking, and recreation. Pier 25 demonstrates how building small parks at frequent intervals along a greenbelt can dynamize the visual and physical landscapes of former industrial wastelands, allowing for people-centered ways of life to unfold along the water's edge.

Pier 40: Community against Big Box Stores

Pier 40 is a large pier complex currently made up of football fields utilized by numerous downtown sports communities. It is one of the few open areas downtown where children can go to run and stretch for any length of space. The story of Pier 40's struggles to derail development plans contrary to the spirit of Manhattan's west side is a remarkable example of how neighborhood activism and engagement saved a historic district from projected out-of-scale development.

Pier 40, a large construction on the Hudson River in a formerly industrial but now largely residential neighborhood with a major highway passing through the vicinity, was being considered as a location for a flagship store for IKEA. The prospect of a big box store in a historic district, whose waterfront was unprotected from zoning manipulations, caused an uproar in the community. Related Companies brought forth another projected development, soon nicknamed "Vegas on the Hudson." Related's plan was to build a Cirque du Soleil theater space with attendant nightlife offerings. Such a prospect would have brought in hordes of new car traffic, burdening the formerly industrial, still-cobblestoned neighborhood and its largely residential character. Active community-board involvement, including children, neighborhood people, parents, commercial enterprises, celebrities, and politicians, generated plans for a mixed-use development proposal, including playspace for children, sports facilities, an art gallery, public areas, and football fields for community use. By early 2013, however, this proposal was foundering, as the Hudson River Park Trust found itself running out of funds to keep its projected expansions afloat.

Pier 40's efforts to create a viable proposal that allows the downtown community access to large amounts of green space that is otherwise unavailable is a test case for how New York will imagine its future. If the city's mandate is to expand its park space, then it has to work with neighborhood coalitions and green-space advocates to find workable solutions for the city's overcrowded dwellers.

Creating green space around New York's circumference today forms a visual complement to the single expanse of green, Central Park, which was considered a mandatory investment in 1853 by its visionary planners, who recognized need when they saw it. Today's New York needs such a visionary leap, something dramatic that affords large amounts of green space where the city's cramped populations can stretch and enjoy their city. Pier 40 continues its battle to remain a place downtown where children and adults can come to exercise and partake in physical activity. How this pier's fate actually unfolds will forecast the lessons New York has yet to learn.

CHAPTER 8

marathon city, biking boroughs

There is something inexplicably impressive about the sight of a densely packed, winding column of bodies disappearing into the horizon on either end of First Avenue, in Manhattan. It is a dramatization of corporeality in New York, a city runner's dream of running in.

Once a year, on the first Sunday in November, New York is a city captured in a snapshot of adrenaline, suspended above water on the Verrazano-Narrows Bridge, framed by a spectacular skyline. The exhilaration pumps through every marathoner. For a moment, a utopian vision of New York City as a human-scale city is created by a gigantic, meandering river of human beings pouring through New York's five boroughs—Staten Island, Brooklyn, Queens, Manhattan, and the Bronx—traversing the world's diasporas in the process.[1]

In 2008, legless athletes in wheelchairs, others armless, sped up First Avenue, followed shortly by a tight pack of female runners flying up the avenue as though it were a hundred-meter sprint. A few minutes later, a colonnade of motorbikes and the ubiquitous yellow-and-black-checked marathon car heralded the first pack of male runners. First Avenue is electric with waves of accolades and whistles. Trumpets and horns, music and motorcade sounds announce the swiftly approaching lead pack of about fifteen men, most of African descent. It is extraordinary, thrilling, raw, impossibly swift. They are gone in the blink of an eye, vanishing against the uptown skyline. For the next six hours, a multitude of runners will course through the city, traveling through all five boroughs, nearly forty thousand in all, attired variously in outfits ranging from marathon couture to post-Halloween accessories, the attire of some runners embossed with slogans and political banners, some festive, other runners seriously running against their best time.

A dynamic and moving enactment of New York's sensibility of big feeling and long duration, the New York City Marathon stages the enormous, manmade environment of Manhattan for the world, framed by Robert Moses's Verrazano-Narrows Bridge and Central Park, designed by Frederick Law Olmsted and Calvert Vaux.

New York City is larger than its mapped, planned, and plotted vectors. Its perspectival axis and great buildings, its monuments and sweeping vistas invoke aspects memorably captured in a photograph, in the mind's eye, or in the graven image. But it is in the compressed, transitory gestures and the provisional dwellings that one captures the fleeting sense of what the metropolis is. Through touch, smell, and the sounds of the city, the larger urban sensorium is indelibly etched as a memory, a taste, a shot of adrenaline, an anticipation, a shudder, an event.

The scale of the annual New York City Marathon is one such merging of dwelling, movement, and event that distinguishes New York City from other urban ecologies. It is the only day the Verrazano-Narrows Bridge is open to pedestrians in an otherwise automobile-choked city. The powerful sensation of a panoramic city filled with people running, walking, and propelling is infectious. The marathon transforms the city of New York into a cordoned-off carnival of feet filling the bridges, the streets, the avenues, and the parkways with a sense of corporeality that is intensely concentrated and epic. It is the festivity of the crowds of Elias Canetti, the modernist writer, without their organized fascistic imperative. It is the random walk of global counterurbanism—a reclaiming of the streets of the globe through running as a technology of movement. The marathon is an invitation to the city to pause and rethink the use of its largest public thoroughfares, its streets and avenues, boulevards and bridges.

During the marathon season, New York City is spectacularly reframed for the pedestrian, the spectator, and the media consumer as a theater for the world. The pulse of the marathoners accentuates an idealized world city filled with the kinetic energy of different arrangements—fast-paced, meandering, loitering, wandering, strolling, standing, hurrying, and of great stillness. New Yorkers experience a provisional ideal of human movement coexisting with nature and technology in a contrived environment. The marathon also highlights the last vestige of nature in New York: the gradient of the ground felt by runners contrasts with the strain of unnatural, manmade slopes, such as the Queensboro Bridge. This theatrical staging of the speeding human form through the jungle of concrete and steel showcases the city of New York as an annual spectacle in which many international anxieties and desires are negotiated in the realm of kinetic urban movement.

A memorable snapshot of the New York City Marathon, the photograph taken in 1980 of Filbert Bayi, a national hero in socialist Tanzania at the height of the Cold War, falling behind, leaving Alberto Salazar, of the United States, to win the gold, captures the spirit of internationalism of the event. Filbert Bayi's running the New York City Marathon, embodying the spirit of New York City for a day, suggests the imaginative power of this particular event in the international imagination. Running through New York is a fantasy for many a marathoner.

Dreaming New York

The connection between New York and East Africa, sparked off by the image of Filbert Bayi, has taken on a physical presence since that famous race in 1980. Since 1980, the New York City Marathons have seen the hegemony of African runners, made formidable by the heightened presence of Kenyan runners. Over 43,000 marathoners were heralded in 2009 by the familiar approach of the lead pack of elite runners, most of whom were of the black diaspora from African states such as Kenya, Tanzania, Ethiopia, Morocco, Angola, and South Africa. Sprinting furiously up First Avenue every year, the stellar first pack of African runners reinscribes older narratives of peoples of the African diaspora who have shaped New York City's public space. A hidden history comes to the fore every year as the newspapers debate a publicly acknowledged nervousness about the predominance of East African runners in the lead pack.

The marathon annually reframes the city of New York as a black, diasporic city, and the city, in turn, reframes the histories of these frequently non-American runners, and their national identities, in terms of the international globalizing of black identities. Most of the established African runners do not reside in New York City, and many of them do not live in the United States. The last few years has seen a slow emergence of recently immigrated East African New Yorkers representing New York City in the lead pack of runners. While most lead black marathoners hail from Africa, they frequently train in Europe, Latin America, the United States, and Mexico, and then run in Berlin, London, Boston, New York, and Osaka, on the route of marathon globetrotters. Their prominence as winners, and, by default, as embodiments of the city of New York, brings to the fore the contradictory dialectic between citizenship and habitation that structures contemporary understandings of urban space.

In New York City, those who most spectacularly embody the hopes and possibilities of this city of extremes are frequently not from the city itself. Instead, they are transient inhabitants like Tegla Loroupe, who trains in Germany, lives in Kenya, and for a day represents the democratic imaginations of New York City, or Adriana Fernández, who became the first Mexican woman to win the marathon, in 1999, or Paul Tergat, who lives in Kenya and who was the favorite for the marathon in 2006. These icons of New York City temporarily inscribe the importance of black diasporic encounters in the promoting of New York City as the globe's replica on an urban scale—the world on an island.

Famous New York City marathoners like John Kagwe (winner in 1997 and 1998); Martin Lel (winner in 2003 and 2007); the world-record holder Paul Tergat (winner in 2005); and, among the women, Tegla Loroupe, the first African woman to win the NYC Marathon and two-time winner of the race; and the two-time World Champion marathoner Catherine Ndereba, all of Kenya, have been New Yorkers for a day. These African runners have variously made international headlines as symbolic embodiments of cities they do not live in: Rotterdam, Paris, London, Berlin, NYC, Osaka.

Their signifying power as world citizens of the marathon circuit and, by extension, as symbols of a heterogeneous and diversified global imagining, is a performance of possibility played against a bleak background of world political disenchantment and the very real politics of race shaping national discourses within the European Union and the United States. As black youth in France and Britain publicly stage their loss of faith in their governments' promise of democratic futures, symbols of black subjectivity, such as the Kenyan Sammy Korir, who holds the world record for Holland's Rotterdam Marathon (2006), and the Kenyan female marathoner Chemokil Chilapong, who won the Belgrade Marathon (2006), promote a provisional sense of inclusiveness and utopian hope to black youth in Europe and the United States, as symbolic representatives of the cities where marathons are held, even if it is only for a day, and primarily in the realm of visual representation.

The image of Tegla Loroupe is a glowing reminder of the importance of the New York City Marathon as an impactful node of convivial internationalism. It foregrounds gender politics, race, and technological virtuosity in constantly changing scenarios. Loroupe's determined image, running before the New York City skyline, her petite figure framed by the soaring backdrop, plays out the dramatic juxtaposition of fantasy and reality, of history and fiction, through which stories emerge from the density of the Marathon.

Techniques of the Body

Writing at the beginning of the twentieth century, the French ethnologist and legendary teacher Marcel Mauss lays the groundwork in his essay "Techniques of the Body" for an interrogation into the nature of movement and physicality. Mauss had visited New York City in 1902 and been hospitalized briefly. He was struck by the pacing and movement of walking in New York City. Writing his essay on the body on his return from the city, Mauss was the first social scientist to study running, gymnastics, diving, and swimming as extensions of socializing techniques through which bodily regimes are trained and managed.

The nature of the physical action, or "techne," for Mauss produces a system of relations between the moving body, the mental action enacted, and the built environment. Techne becomes a fundamental suturing device between being and becoming, between the self and the world. Mauss suggests that styles of running, philosophies regarding the posture of the body while running, and the particularity of body management produced by the act of running (involving discipline, the diet, the training regimen, and choreography) offer an architectonics of personhood. The history of a particular techne is the history of a given moment and spatial configuration, enacted within a narrative of limits. The runner runs against the available limits of her time, in speed, distance, and physical possibilities. Running becomes a way of knowing the city.

In the marathon, demands are made on the eye; the ear offers a different register of reading the event, while the spectator watches with an alert yet blasé curiosity, lending narrative to the unfolding physical structures of moving bodies, waves of sound, and the spectacular receding distance of the horizon.

For the spectator, the focus is on process, rather than on the denouement. One must actively seek a good spot, move to catch the runners at the next strategic watch point, rush by subway to intercept the lead pack, or stay at one point and watch the six-hour-long trail of human bodies wind their way through the geography of the city.

Mauss's phenomenology of technique argues that the minutiae of movement offer an elaborate system of signs through which modernity is produced. Modern subjectivity is a process of physical reinvention and bodily renegotiation—a sloughing-off of older regimes of body management and a taking on of new forms that are ever being advertised, imposed, advocated, forms that seduce us and surreptitiously envelop our everyday life. Forms of work, leisure, socialization, boredom, and entertainment produce changing notions of techne, as well as the speed by which such actions ought to be realized.

According to Mauss's theory, the act of running is a technology of the self with far-reaching social implications. The predominance of black diasporic marathoners has shifted the politics of running internationally. Once an innocent activity done privately, running has become a more abstract project in global becoming. Running, through the images of black marathoners representing world cities like London, Paris, New York, and Osaka, foregrounds the social hope for democratic transformations on a local level, across national contexts.

Running through the City

Tegla Loroupe's petite figure emerges as a powerful icon in the New York marathon. Under five feet and weighing only eighty-four pounds, she is the first African woman in the marathon's history to set a World Marathon best. Loroupe won the New York City Marathon twice, in 1994 and 1995, and was three-time World Half-Marathon Champion, two-time world record holder in the marathon, and the Kenyan national-record holder at three thousand meters, five thousand meters, and ten thousand meters.

Loroupe's compact physique tenaciously hugs the burning asphalt. Like the relentless verticality of the city, hers is a moving frame of muscularity and monumentality. Fast-paced bodies pound the length of road. The theatricality of the street and virtuosity of running converge in crescendos of applause and elation, exploding in wave after wave, corner after corner, in a tumultuous cascade of sound and bodies.

In a stunning marathon race in 1998, Loroupe sprinted up First Avenue in tight tandem with the NYC Marathon winner of 1997, Franziska Rochat-Moser, of Switzerland, and with Franca Fiacconi, of Italy, Adriana Fernández, of Mexico, and Ludmila Petrova, of Russia. It was a swift and powerful staging of the poetics of grace, speed, and verticality. The image was speedily devoured by the impressive moving curtain of human forms fleshing up the networks of transportation with a scale of motion that brought the city into a trancelike state of spirited disruption and persistent hope.

The muscularity of motion meets the spirit of challenge, as constellations of human bodies pass by in attitudes of exhaustion, tenacity, elation, delirium, and concentration. Human imagination and adrenalin confront hardscape. The density of crowds in motion, contrasting with the built environment of New York, hypnotizes one into a state of exhilarated spectatorship.

Now devoured by the gargantuan snaking form of runners pulsing through the spectacular Verrazano-Narrows Bridge, now etched against the winding neighborhoods of Queens, Brooklyn, and Manhattan, Loroupe's slim figure in the lead pack of female runners offered a narrative of migrancy inscribed on concrete, a trace of spirit embracing the entrails of the city, hugging its contours, carved out by Robert Moses, and elaborated by everyday practice.

At once transient and immemorially stamped on the amnesiac macadam, Loroupe's poignant journey of passion forced into conversation other accidental encounters of civic exchange, urban cordiality, and the shared anticipation of a humanist event in an age of inalienable difference. A city brutally rendered asunder for a culture of cars is momentarily humanized, retracing its memory of other kinds of interaction through this mobile thread of runners who throb through the city's layers of history, layers built on the silent and repressed history of black New Yorkers' labor.

The epic narrative structure of the New York Marathon disrupts notions of winning and losing to open up ways of sensing the city as an experience to be run. To participate, not to win, makes the point of the running a pleasurable epic of little stories unfolding. Human perseverance and public spectacle combine to create an irrational theater of endurance and strength. Each episode, each of the 43,000 passing runners, reveals a different tale on the nature of running and becoming. Yet what frames this extraordinary internationalized performance of urban endurance is the stunning visual power of black diasporic identities contesting for the imaginative power wrapped up in the symbolic image of representing New York City.

The particular embodiment of Tegla Loroupe, who in the 1990s was considered the fastest woman in the world and the greatest female marathoner of all time, staged the technology of running as a technique of self-determination on a global platform, using her social consciousness as leading figure for world peace. She has now furthered her international presence through her organization, the Tegla Loroupe Peace Foundation, and has thus repositioned the importance of the New York City Marathon as a site for progressive internationalism in new ways.

Biking across Boroughs

On Friday, October 29, 2004, ghouls, fantastical beings, supernatural spirits, demonic creatures, activists, and anarchists—cycling enthusiasts all—crowded the piazza on the north side of Union Square for the seventh cross-Manhattan bike ride, held on Halloween weekend. The event is a spontaneous gathering of urban bicyclists called Critical Mass, who cycle through three hundred cities across the world, without permits, reclaiming their cities' roads for low-tech commuting.

Critical Mass is an ecologically committed monthly cycling event. There is no single individual in New York City who claims to be the head of the organization. It was initiated in 1992 in San Francisco and swiftly gained popularity in cities across the United States and in other countries. The principle is to gather at seven p.m. on the last Friday of every month and cycle through the city's streets with no prearranged route. Implicit in the improvised nature of Critical Mass are spontaneous and multiple confrontations with cars and traffic rules, entities that all cyclists must navigate. Such a flash-mob-improvised mobilization of cyclists hurtling through the streets of major cities draws attention to the marginality of the cyclist as urban entity and highlights the tenuous nature of the cyclist's right to the city's macadam pathways.

The popularity of the monthly event in over three hundred environmentally engaged cities around the world gives the activity a global political relevance. Cycling becomes an open statement against fuel consumption and unsustainable modernity. It is an active and vigilant engagement that challenges the privileging of car culture over other forms of transportation—a Jane Jacobs human-scale philosophy of the city that is opposed to the Robert Moses logic of a car city.

Inadequate levies and taxes on the use of vehicles by those who do not live on Manhattan Island have made the island's roadways unmanageable and its quality of life a major issue for its inhabitants, Manhattan's residents would argue. Critical Mass emphasizes human-scale transportation and the search for alternative modes of travel to replace oil-dependent varieties.

The mood during Critical Mass events is more like that of a carnival or celebration of urban mobility than a straightforward statement against car culture, per se. Many participants dress festively in dramatic costumes. The types of non-oil-based two-wheelers ridden by participants are as diverse and idiosyncratic as the characters who join in the two-hour ride: tandem bicycles, rollerblades, babies in perambulators, acrobats on stilt bicycles, people in wheelchairs, homemade cycle rickshaws, a man with a dolly loaded with cardboard boxes, and other inventive contraptions, such as horizontal-pedaled low-slung bikes and go-carts.

The informal group does not claim to be an activist group; but its philosophy stems from a desire to promote an alternative consciousness. Its philosophical impetus is to assert the communal right to the street and the highway. One goal is to prioritize alternative forms of transportation over the oil-determined options currently privileged. The youthful and defiant energy of Critical Mass, since 1993, has transformed the festive event from one of a pleasurable reclaiming of the highway and street for the bicyclist and the skater to a tense and public struggle between city policing statutes, the car population, and the increasingly militant bicycling advocates for the right to parade without permits through city streets.[2]

The group's strategies in New York City have been to surprise the city, as its spontaneous human-powered eruptions into the flow of traffic are mobile in ways that the car is not. Tactics include riding against the flow of traffic, moving in directions contrary to traffic rules, and breaking off into spontaneous subgroups, thus freeing up the circulation of the city for non-fuel-based transportation systems. This unpredictability has historical roots in an anarchist history of urban self-invention whose antecedents are the European Situationists, rather than Robert Moses.

The Halloween Critical Mass Ride of November 2004 was particularly tense, due to the grinding disappointment with the November elections that voted George Bush into office for a second term. On October 29, 2004, the City of New York brought an injunction against Critical Mass to prevent cycling en masse through the city. The injunction was denied, and the November event thereby acquired a strong new prominence and import in the public's consciousness. Critical Mass became the site of the right to due process and the First Amendment right to free association.[3]

The increasingly policed state of New York City's public culture accrued attention around the debates and issues staged by Critical Mass, which wanted the right to cycle freely around the city, without a permit or being surveilled. The police department had distributed flyers at the starting point of the bike rally, laying out the rules under which they would arrest people, despite the recent judicial decision that gave the bikers the right to cycle through the city without a permit. The amassing of police vehicles on the side streets of Union Square and the suffusion of police into the main thoroughfares reminded participants of the recent dragnet-sweep at the Republican National Convention, strategies employed by the police to arrest nearly 1,500 protestors and bystanders.[4] Barking out the conditions of mobility, which included a police-sanctioned route for the ride, the city police once again found themselves at odds with the people of the city. It was only a matter of time before the tense conditions of order would crumble.

Scores of specters haunted New York City streets that Halloween, apart from the usual panoply of fantastical beings. Enough Osama bin Ladens, Arafats, Idi Amins, and people in khafiyas and headscarves paraded through the streets of Manhattan to remind the city of the real hauntings pursuing the American psyche at the time. At Union Square, the mood was anger at the police and anxiety about the country's future. "Repeal the First Amendment," read placards held by the top-hat-and-black-tie-attired "Billionaires for Bush" activists. "Again we ride," read the T-shirts of other Critical Mass enthusiasts. "No oil for war," read yet another placard. Superhumans, celestial beings, grotesque creatures, bunny rabbits, bestial monsters, and a cardboard walking Hummer conglomerated in the carnivalesque air electric with surveillance, the police presence, and hovering helicopters.

The thousand-strong entourage departed from Union Square with great fanfare, hooting and screeching, celebrating and hoorahing, as the event began spontaneously. They sped up Park Avenue, exuding a sense of possibility and freedom seldom experienced in a car-determined city. The specified route involved going up Park Avenue to Fifty-Fifth Street and then taking a left to return down Broadway, ending up at Union Square in a highly surveilled and monitored route. Needless to say, the predetermined route conflicted with the philosophy of the group, which believed it had the right to cycle peacefully through the city in any manner it desired, taking any route spontaneously. This is precisely what ensued in what turned out to be a nerve-racking evening of spiraling helicopters, police sirens, and reveling riders taking the city back from the police presence. Cycling for over an hour and a half, the group soon broke up into various subgroups, spontaneously diverging at different points, heading up to Times Square, up Eighth Avenue, down the West Side Highway, and across Manhattan. The city was being transformed, even as paranoid viewers and recalcitrant police watched with mixed feelings.

The sense of taking back the city, even for a moment, and envisioning other ways of being in the city's main thoroughfares, was thrilling. Dramatically attired beings in tinsel and glitter sped by on tandem bikes, architectural bikes, clown bikes, and skates. The air was thick with tension, but youthful with hope and spirit, as the spirit of New York City was staged by those still daring enough to have vision in an era of closed perspectives and diminished hope. At the end of the electric evening of contentious energy and alert police, thirty-three arrests were made, primarily for violating the prescribed route the police had sanctioned.

Green Mobility, Activist Space, Memorials

On the last Friday of September 2007, musicians, drummers, dancers, trumpeters, hula-hoopers, and bikers filled the north piazza of Union Square in the fading evening light as organic farmers who attended the Union Square Farmers Market wrapped up their day's labors. The air was festive with lively big-band circus music, enticing passersby into a jig, if not a conversation about greening the city through green transportation.

Improvised, daring, imaginative, innovative, inspirational, mobile, and global, the impact of Critical Mass, as a New York phenomenon, in concert with bicyclists around the globe, has been far-reaching. The choice of the term "critical mass" evokes Jane Jacobs's advocacy for city planning that allows for accessibility and a wide range of encounters within cities, by considering the density of human movement and transportation. Jacobs's far-sighted commitment to creating a sustainable urban environment was never fully accepted as an efficient way of thinking about energy and sustainability within cities, during her time in New York City in the 1960s. She envisioned mixed land use, complemented by a varied transportation system, one made up of subways, buses, cyclists, and people who could walk to their grocery stores, but a system that also accommodates the influx of cars that interconnect the five boroughs.

The environmentally engaged encounters of Critical Mass with the water-bound ecology of the city of New York encourage the use of alternative energy sources for transportation and the compelling arguments for a safer pedestrian-defined network of transportation systems to be continuously explored within New York City. The most recent conversation centers around the allocating of tolls for cars entering Manhattan, in order to curb the massive influx of single-car commuters and unsustainable transportation models, such as the heavily designed SUVs and family vans that do not fit the scale of the streets of downtown New York City. Accompanying this conversation is the increasing visibility of cyclist memorials around the city.

Ghost Bikes

A white bicycle with yellow daisies lies tethered to a pole at Thirty-Sixth Street and Fifth Avenue. A garland of fresh flowers encircles it, framing a simple cardboard memorial. On the West Side, on the Hudson River Park bike path, two white bicycles with memorial plaques and garlands are fastened to posts. One carries a handwritten note in black ink that says, "With Love and Rage." On Houston Street, near Allen Street, a ghost bike remembers a young soul with "Rest in Peace." It is ceremonially tended as a grave, with dedications and offerings. White paint, colorful wreaths, and placards of grief mark these ghostly public memorials. Ghost bikes remind the passerby of the safety and mobility of the urban dweller. They arouse the stranger's melancholia as he encounters a site of violent death.

Summer Streets

Critical Mass's fraught encounters with New York City during the 1990s led to a growing awareness of bicycling as a popular mode of transportation in New York. The impact of bicycle advocacy groups like Time's Up and Transportation Alternatives, and far-sighted bicycle planning across North America, in cities like Toronto, Seattle, Chicago, Pasadena, Madison, and Portland, has generated a growing consciousness in New York of the potential of urban cycling.

Drawing on this momentum, Mayor Bloomberg, in his second term, attempted a rethinking of New York City's traffic flows by raising the thorny topic of congestion pricing, a proposal to charge vehicles a toll for driving into Manhattan during peak hours. Bloomberg's unsuccessful efforts to implement congestion pricing in early 2008 were followed by a slight decrease in traffic into Manhattan as a result of escalating oil and gas prices. What policy could not achieve, the global oil crisis did, reducing congestion on a daily basis in New York City.

Under pressure from advocacy groups, such as Transportation Alternatives, to further experiment in opening up the city's streets to alternative modes of transportation, the Department of Transportation initiated a program called Summer Streets, in August 2008.

Summer Streets included three Saturdays in August, when all of Park Avenue, from Central Park down to the Brooklyn Bridge, would be free of traffic until one o'clock p.m. This project incorporated bike-share arrangements, whereby people could rent bicycles and return them. Free skate- and bike-repair facilities, free yoga, exercise, and dance workshops, and tents with drinks and refreshments, along an otherwise impossibly busy avenue, transformed Park Avenue into an extraordinary provisional public park.

Strollers, bikers, skaters, dancers, wheelchairs, runners, scooters, skateboarders, walkers, and people on beach chairs reclaimed the streets of the city as a place of enjoyment, leisure, and shared urban peace. The city was out on the streets in every shape, form, and movement: soccer players, baseball players, tennis players, and jump-ropers. The side streets, blocked to traffic, became enclaves of human interaction. Cycling under the Pan Am Building, now the Met Life Building, New Yorkers marveled at the unique sensation of cycling and walking over the great Grand Central Terminal. Sensually lingering on the bridge over it, throngs curved into the great Met Life Building arches, its thoroughfares normally clogged with car traffic. A sense of tremendous excitement, elation, wonder, and public delight permeated this great gift of the city to its inhabitants.

Summer Streets 2009

Picture Park Avenue South's magisterial approach to Grand Central Station, filled with human movement along its broad avenue. Picture the view from Grand Central, looking uptown, all human-scale movement in different states of propulsion. The image you conjure is the impossible scenario of Manhattan without cars, or at least a portion of the center of Manhattan—Park Avenue, Lexington, and Madison—without cars.

Introducing urban dwellers to a day without cars, at least a few hours without vehicular transportation, Summer Streets forces city people into new relationships with their environment. The absence of cars creates a clarity in the city's movement flows.

Summer Streets began as an incentive for New Yorkers to keep the dialogue on congestion pricing alive. What if the city looked toward alternative transportation networks other than cars for primary commutes within the city? Rearranging individual relationships to traveling in the city and dependency on the car allows a recalibration of other issues as well: those of health, energy consumption, social encounter, speed, quality of life, air quality, pollution, noise levels, stress reduction, and increased longevity through exercise. All of these benefits make the scenario of a green beltway that offers alternative transportation networks increasingly attractive to New Yorkers.

Since the Hudson River Park Trust completed a large part of its greenway on the west side of downtown Manhattan, New Yorkers have increasingly become accustomed to cycling, instead of taking other transport to work and play. By 2013, the presence of commuter bicyclists had exploded. The ongoing development of Riverside Park, and further along the Upper West Side, expands this beltway all the way to the Bronx. It is now possible to cycle across all the boroughs, excluding Staten Island.

Under Mayor Bloomberg's administration, the commissioner of the New York City Department of Transportation, Janette Sadik-Khan, exponentially expanded New York's quest for a sustainable future of fluid potential. Under Sadik-Khan's often controversial but persistently innovative pilot experiments, New York has learned from other eco-efficient cities around the world. From Copenhagen, the idea of street calming was introduced. From Amsterdam, the idea of free bike-sharing has influenced New York. From London, the idea of congestion pricing was introduced as a potential energy-saving legislation, with unsuccessful results. From Curitiba, Brazil, the notion of a rapid bus system is being implemented in New York. From Berkeley, Portland, and Palo Alto, the idea of parallel bike routes were introduced in New York City. Mayor Bloomberg's call to learn from Copenhagen has expanded to learning from Amsterdam, London, American cities, and most recently, Curitiba, as New York City pursues eco-efficiency with a priority on sustainability and bottom-up involvement by commuters and the city itself.

Learning from New York

"New York City is the bicycling capital of the world," according to Janette Sadik-Khan. The culture of bicycling has opened up entirely new ways of experiencing New York City. It is slowly transforming the city into a network of bicycle commuters. From the fraught encounters between Critical Mass and the police, to the expanding culture of biking as a way of life in

New York today, the story of bicycling in New York serves as an instance of the city's denizens exploring its utopian possibilities.

New York's energy-saving efforts have been bolstered by visionary and experimental planning on the part of the Department of Transportation and the New York City Department of Health, as well as the New York City Parks and Recreation Department and independent borough leaders, such as the Bronx borough president Rubén Díaz Jr.

From 2006 to 2011, the city saw the rapid installation of bicycle networks in the South Bronx, Long Island, Williamsburg, downtown Brooklyn, and downtown Manhattan, between the East River and the Hudson River. During this period, over two hundred miles of bike lanes were completed, generating what amounts to a bike lane from New York City to Boston, according to the Department of Transportation's press release.[5]

A noteworthy achievement is the creation of access to the four East River bridges, with designated bike lanes, in particular the emergence of the protected bicycle lanes, situating bicyclists between the curb and the parking lane, a planning innovation that has received considerable public acclaim. This experiment, whose first site was on Ninth Avenue, has increased the number of cycle commuters on New York's streets. Its originality has drawn attention to the cyclist as an important street entity. According to the Department of Transportation, the protected bicycle lanes created in New York City are the first of their kind in the United States.

After lagging behind many American cities in its greening initiatives, New York City has finally made serious inroads into improving its commuting accessibility. Its multipronged approach to innovative street planning, the creation of green bicycle lanes, the designing of bicycle boxes on streets, the experiment of protected bike lanes, the construction of designer bike racks, and establishing attractive bicycle guide signs at major bicycle intersections has encouraged a new engagement with New York City's hardscape. Cycling over the Brooklyn Bridge on my way to work in Brooklyn is an extraordinary spectacle every day, revealing new details about the city unavailable at street level. The bicycle revolution has opened up New York's intimidating skyline as a human adventure to be pursued by those lured by its spectacular vistas.

New York City's Department of Health has visibly stepped up its interest in promoting bicycling as a citywide activity. "Regular physical activity not only helps prevent heart disease but also reduces your risk of diabetes, colon cancer, breast cancer and depression," observes New York City health commissioner Thomas Farley.[6] Incentives used to promote the new culture of bicycling to work in New York City include recreational pleasure, an environmentally compatible commuting plan, economic benefits, and the added advantage of exercise. This new relationship with the street, the citizen, and the hardscape is still in formation.

New York's bicycle lanes are not safe enough yet. Drivers of cars do not observe bicycle lanes with the serious attention they merit, and the sense of vulnerability of daily bicycle commuters has not receded. Cyclists are very much aware that their newfound mobility and access to the city is a foolhardy tryst with the vagaries of New York's traffic, given the fundamental disrespect for cyclists. However, the culture of bicycling to work is here to stay. More people are taking to the streets on their cycles.

Critical momentum will slowly create a deeper awareness of New York's cyclist as a respectable road entity. In this respect, New York's traffic still needs to learn from that in Copenhagen. The inroads into cycling that New York City has made from 2006 to 2013 has won it accolades as one of the most "improved" cites for cycling in the United States. It has been recognized by the League of American Bicyclists as a cycling-friendly community—the only city on the East Coast to receive this recognition.[7] In this respect, New York City is a mirror for the future of other American cities that have yet to open their streets to the bicycle.

maritime mentalities

New York City is enmeshed in a new maritime sensibility. This twenty-first-century perception is not driven by global trade and conquest, but by a studied realization of one of the city's own strengths: its waterfront. It is an embrace of the city's cosmopolitan roots as an immigrant city of arrival, a city founded on the belief that commerce is the only religion.

The mercantile foundations of this settler trading city positions its history of secular cosmopolitanism in contrast to its counterparts elsewhere in the Dutch East India Company's possessions. National claims to myths of origin produced different renderings of cosmopolitan potentiality in other colonial port cities. In seventeenth-century New York, every arrival, whether immigrant, refugee, traveller, or slave, technically had rights to the city, as the island's original inhabitants had been dispossessed of their rights through genocide. However, that potential was not immediately realizable, in legal terms. The gap between the promise of individual freedom and the competing restrictions bracketing rights to the city for women, slaves, and, momentarily, Jews, to name a few, captures the conflicted interests shaping the emerging political landscape of cosmopolitan New York.

Despite the restrictions imposed on the "open" Dutch city by the colonizing British presence in 1664, an international, heteroglot culture of competing opinions and lifestyles took hold on the streets of New York, with its piers, slips, and waterfronts acquiring an intensity of political ferment that was fraught with anticolonial sentiments, as it was vibrant in its youthful embrace of modernity.

Most distinctively, however, New York's maritime cosmopolitanism was founded on the empirical fact of its being a colonial city, created through what Kant identifies as "un-cosmopolitan" means. Writing in the eighteenth century, Kant observes that colonial conquest displays the abnegation of the code of hospitality, of visitation, as it implied a usurpation, a forced occupancy. Following Kant's logic, the colonial city, bereft of a culture of uncoerced urban life, cannot offer the potential for individual freedom to its urban dwellers. Furthermore, a city's culture of hospitality contains within it a notion of "return" of some sort, for the guest. British imperialism lay in store for colonial New York after the Dutch no longer owned Mannahatta, bought for a sum of sixty guilders. This double negation of the narrative of hospitality extended a further violation to an unsustainable act, that of expansionist occupation on foreign soil.

The disassociation of early New York from a narrative of "return," a return the Dutch were offered by the British in 1664, and which the Dutch declined without much pause, presents a peculiar twist on the Enlightenment discourse of hospitality. It bears a conditionality of openness that omits a literal leaving.

Historically, if you had immigrated to New York, you were assumed to have made a one-way passage. There was no return embedded in the idea of arrival. New York was too far away from Europe, and the trip too expensive, too dangerous, to allow the immigrant to envisage a return. To that effect, the challenge to be successful in New York was all the more dire. There was nowhere else to go, if life in New York was not working out. Early New York cosmopolitanism was, to that extent, a tough-love cosmopolitanism, necessitating mutual coexistence for the trader, the traveler, the sojourner, and the slave.

Today's New York bears the self-conscious realization that this is a city with a memory of colonial coercion and a history of independent thinking. Its waterfronts bear the trace of its maritime expansiveness, through recurring public rituals, such as carnivals, festivities, and civic commemorations. This final section presents a varied picture of events that capture the city's maritime heritage while also furthering the city's commitment to cosmopolitan urban belonging. The city pursues mechanisms whereby democratic social engagement furthers the city's public responsibility to build a just, equitable, and sustainable metropolis.

CHAPTER 9

brooklyn carnival and the sale of dreamland

Every summer and fall, the city of New York embraces the liberties of heat and lightweight clothing. The summer of 2008, in particular, stands out in my memory as a summer when pedestrians consciously reclaimed the streets of New York. Long, festive, nondenominational celebrations of people reveling in the streets on designated car-free days, taking up spaces in certain areas of New York City normally assigned to cars and heavy traffic, focused attention on a critical aspect of life in the city, that of cosmopolitan identity. This theatricalization of cosmopolitan citizenship is usually available only during specific large-scale events, such as parades and carnivals, of which the Brooklyn Carnival is the most elaborate.

The experience of cosmopolitanism is enmeshed with the geography of a city. New York City's peculiar landscape of islands and peninsulas interlocks an imagined unit of five entwined boroughs. It produces urban identities that are as fragmented and diverse in their local imaginings as neighborhoods and communities are varied.

New York's structures of imagining are perpetually in emergence. They draw upon multiple transportation histories. Green swaths thread into the water-defined pathways of this old port city to produce changing perceptions of its waterfronts, piers, bridges, and canals. From the maritime waterways of an earlier era of global seafaring economies, to its numerous postnationalist, annual public gatherings in the form of parades, protests, carnivals, and festivals, New York's mental image of itself is shaped by airways, waterways, railways, roadways, and now its digitally networked pathways. Passengers on ships, planes, trains, and the region's roadways merge to create New York's sense of itself as a cosmopolitan crucible whose way of life demands bodily and social intimacies that are particular to dense, networked living.

One locus of such cosmopolitan intimacies is the neighborhood carnival, several of which have grown to the status of non-nationalistic parades, with all their attendant requirements, permits, surveillance, barricades, policing, contained routes, and a predetermined time of beginning and completion. The Mermaid Parade, in Coney Island, the Gay Pride Parade, the Halloween parade in Greenwich Village, the Earth Day Parade, and the New York City Marathon are such loci of urban intimacies. Their very different pathways are rooted in New York City's mythic relationship to particular localities, such as the New York waterfront, the pagan ritual of All Saints Day, rooted in Greenwich Village's own location as a burial ground, and the citywide commitment to human-scale activity most popularly embodied in the New York City Marathon.

Of these myriad, spectacular metropolitan public events, the Brooklyn Carnival, which unfurls on Eastern Parkway every Labor Day, displays the most dramatic blurring of race, nationality, ethnicity, and notions of cosmopolitan civic imagining. Beginning with the disrupted early morning train service, shuttling three million people from all directions to Eastern Parkway, the Brooklyn Carnival is firmly embedded in the bedrock of New York's daily economy. A range of civic and mercantile organizations, as well as a broad spectrum of New York's cultural base, is found in Brooklyn. Spanning the Caribbean and Central and South America, the festival showcases the elaborate interfaces between long-distance nationalisms and the American melting pot.[1]

On that temperamental, rainy Labor Day in 2008, children in gold lamé and colorful regalia, excitedly clamoring against beach chairs on the standing-only subway car, spark the excitement of New York at the height of summer. It is New York at its most hopeful. On this one day of the year, West Indians, Latinos, Asians, and Anglo-Americans blur into an Atlantic cultural fusion. The air is aromatic with Worcestershire sauce, jerk chicken, saltfish, codfish, callaloo, sorrel juice, sea moss, grilled fish, and goat meat, startling one into realizing the intermeshing of cultures and cuisines amid the brownstones and the beaux-arts buildings, the sweeping boulevards, and the expansive Grand Army Plaza and Franklin Avenue areas.

The parade of 2008 boasted floats and displays by groups such as the Fire Department of New York, the New York City Department of Correction Law Enforcement Explorers, the Association of Caribbean Americans in Correction, the National Association of Professional Firefighters, numerous religious congregations, including the Haitian Apostolate, the Diocese of Brooklyn, the Church of St. Benedict Joseph Labre, the Church of St. Gregory, Our Lady of the Presentation Church, and other community organizations.

The Brooklyn Carnival is different every year in its ambience and impact. Often shrouded by an unpredictable August shower, which has frequently dampened a vibrant, full-blooded throng of revelers, the parade is a microcosm of what the city has been through in the past year. The parade demonstrates the powerful civic expressivity of immigrant communities and diasporic organizations, celebrating New Yorkers through live and canned music of the Caribbean diaspora, street dancing as well as motorized floats of civic entities, foods from the Caribbean, regional ethnic commodities, and fantastical costumes that rival the more world-renowned carnival cultures of Rio de Janeiro and Trinidad.

The festive, effusive staging of cosmopolitan conviviality allows the workaday realities of immigrants and their second- and third-generation American relatives a joyous respite that is firmly rooted in notions of labor, work ethic, and the iconicity of New York City.

The Brooklyn Carnival acquires a cosmopolitanism particular to New York, with its material and historic links to the West Indian communities in Brooklyn, as well as its diasporic black communities in Harlem, Queens, and East New York. The event always promises the biggest party on either side of the Atlantic, linking the carnivals of Bahia, Port-au-Prince, Kingston, Trinidad, and Panjim to that of New York.

Urban Space as Carnival

The Brooklyn Carnival is a highly political conduit for the public displays and cultural negotiations that layer cosmopolitan citizenship. While the occasion for carnival is one of the oldest expressions of an urban civic ideal, the idea of carnival in the twenty-first century has given rise to complex manifestations of cosmopolitan belonging. Carnival has emerged in the urban cultures of the North as an intensely physical, public articulation of difference and coexistence in the city.

Carnival is not merely an event for celebration, nor is it simply a vehicle for playfully destabilizing authority. For many displaced and migrant communities, carnival has come to embody a crucible of psychic negotiation between the city and its new immigrant populace. It physically permeates the geography of the city's imaginary, revealing the often tense demographic transformations that make up the city's sense of itself. Carnival persists beyond the actual parade and cordoned-off social space for ludic social interaction. It is a way of being in the city that stages a distended cosmopolitanism, resounding with echoes of Trinidad, Britain, Brazil, and Goa.

This idea of carnival has its antecedents in the socially sanctioned urban playfulness of the Middle Ages and the Renaissance, playfulness that gave rise to the jester, the village fool, and the clown. But carnival's implications for postcolonial subjectivity in metropolitan centers are specific. In the contemporary city, the carnival is both an emancipatory social space and a pathologizing sentiment that determines how specific bodies are read within the urban.

Carnivals in the United States, Britain, and Canada contain different political stakes for Third World immigrants when compared to carnivals in Third World states. At the Brooklyn carnival, the attendees' presence as citizens and pleasure-imbibing subjects asserts their civic and legal legitimacy. In this reworking of postcolonial ontology, carnival is the self-aware condition of coexisting in the city of policed difference.

Modern carnival is structured as a carefully contained map of parades and civic expressions. Its elaborate costumes, music, themes, colors, and hedonistic mood of celebration often reinforce preexisting stereotypes about particular communities. An influx of commuter participants complicates the demographics of such urban crowd events. To that extent, carnival is always a translocal embrace of citizens with their city. It is an expression of the city's laws, regimes, and discourses, as well as its citizens' negotiation of urban limits and possibilities.

The history of minority urban citizenship in metropolitan centers is illuminated through the interlocking sites of the Brooklyn Carnival, the Nottinghill Carnival, in London, the Brazilian Carnival, in Rio de Janeiro, and the Trinidadian Carnival, all of which highlight the struggles for civic identity encapsulated in the notion of carnival as an event for immigrant communities in the contemporary city.

The Brooklyn Carnival is particularly emphatic about its civic investments in New York City. Its role as an annual performative event evokes the history of the city's globally drawn black citizenry. Through the exchange and display of music, food, civic presentations, and public representation, the event demystifies the production of black New York identities in contradistinction to the pathologizing discourse of immigrants as delinquent and violent. Furthermore, it accentuates the tangible creation of a distended cosmopolitanism that is grounded in New York, but draws its traces, etchings, and echoes from other localities, such as New Jersey; other cultures, such as that of Haiti; and other nations, such as the countries of the West Indies.

Transnational Civic Identities and Urban Citizenship

In the United States, the prominence of the Brazilian and Trinidadian carnivals has tended to overshadow the importance of the Brooklyn Carnival to the interconnected, nomadic itineraries of carnivals across the world. Ritualistic performances of both the Brazilian and Trinidadian traditions inflect the Brooklyn Carnival with ideas of public spectacle, the production of the marvelous, and notions of ritual embodiment in public space.

The Trinidadian Carnival has often been invoked to exoticize and pathologize black citizenship in Britain. Coming out of the history of the Nottinghill Carnival, in Britain, and its relationship to British racism, and the multiple histories of African, Caribbean, and British traditions of carnival and resistance, black British cultural practices have self-consciously addressed questions of history and public spectacle through the Nottinghill and Caribbean carnival traditions.[2]

In Britain, media representations of carnival, as in the BBC's reporting, focus on the ethnographic, rather than performative, nature of the spectacle itself. British reporting of carnival tends to relegate its derivative origin to the Caribbean, eliding the everyday, labor-based roots of the Nottinghill Carnival in London's West Indian community. Research on the Nottinghill Carnival suggests that British media distorts local political formations and struggles for British citizenship contained within the Nottinghill Carnival's statement as an urban, public event, and its claims to be a civic expression, rather than an event of "unrest." The Nottinghill Carnival is frequently portrayed as a reverberation of the Trinidadian Carnival, and not an event with roots in the political and cultural geography of London.

Brooklyn's Labor Day Parade distinguishes itself from other carnivals by grounding its impetus in the politics of labor itself. The carnival becomes an opportunity to invent public memory by revisiting histories of recent migrations and former nations of belonging, all condensed into concurrent struggles for local legitimacy in New York City itself.

The City and the Citizen

In 1997, the spectacle of the West Indian Carnival in Brooklyn was particularly moving, occurring as it did on the heels of Abner Louima's torture and disablement by the New York City Police Department. On August 9, 1997, the thirty-year-old Haitian immigrant and father was mistakenly arrested in a Brooklyn nightclub, violently assaulted, and sodomized with a toilet plunger by police officers from the Seventieth Precinct, in Flatbush, Brooklyn. The horror of the case wrapped New York City in shadows for weeks.

Louima's case broke the tense silence hovering around the issue of policing and brutality. Here was a typical middle-class New Yorker: a young, downwardly mobile immigrant whose life one night came violently undone in a devastating encounter between law and life. Louima was a Creole-speaking Haitian from Port-au-Prince. He trained as an electrical engineer in Haiti, but worked as a security guard in New York. The Louima incident spotlighted a number of issues regarding citizenship in New York. It showed the city as an immigrant city, where English is only one of the many languages through which civic identity is expressed, as Louima was interviewed in Creole. Louima's horrifying ordeal held the city in a petrified stupor, as the toilet plunger, the instrument of Louima's torture, became emblematic of the city's sewers of failed justice and blocked arteries of civility and civil rights. At once banal and sinister, the plunger became a symptom of the city's moribund and decaying justice system, its excremental policing practices. As a symbol, the plunger accrued a visual power of unprecedented proportions, at once a costume of carnival and a statement of solidarity with other concerned citizens in the city.

For the Haitian community in New York, Abner Louima's tale forced open hard issues that middle-class aspirations had prevented them from confronting in politically organized ways. The scandal, according to many Haitian immigrants, compelled a new awareness of racism and their own vulnerability. Many Haitian immigrants had fled a dictatorship in their homeland for close-knit communities in New York and other cities. Louima's story forced them to recognize the need for more public investment in questions of urban and civic citizenship in the United States. The incident became a fulcrum for turning the Haitian community toward an issue it had never been committed to before.

Ricot Dupuy, the general manager of Radio Soleil d'Haiti, in New York, summed up the situation in this way: "Haitian organizational structures are definitely lacking. There was so much wrong in Haiti that Haitian [immigrants] were forced to direct their energy toward changing it, so they were Haiti-focused rather than U.S.-focused. Now they realize they are an integral part of U.S. society."[3] As Dupuy's comments suggest, the ambivalence of this diasporic community toward accepting the role of politically engaged citizens with civic commitments to New York, and their nostalgic, sojourner sense of place in Haiti, was called into question through this violent event.

The questions raised by the Louima incident, regarding the Haitian community's political efficaciousness in New York, as well as the community's own unresolved relationship to race and minority identity as an immigrant community in the United States, resonate with other immigrant groups as well. The case highlighted a familiar tension between the predetermined illegitimacy of diasporic communities—despite their presence as an enterprising constituency—in their current country of domicile and their attachments to former countries of imagined community. This interlocking condition raises broader issues about the nature of cultural and civic citizenship and possible avenues for new and changing political imaginaries that shape locally derived social geographies and emerging notions of the civic.

Sites and Civic Ordinances

Three weeks after the Louima incident, on the heels of a major protest march held August 29, 1997, the "second largest party in North America" unfolded as a public expression of urban citizenship. The press releases announcing the world's largest neighborhood block party—starting on Eastern Parkway and Utica Avenue, continuing west on Eastern Parkway to Grand Army Plaza, then on to Flatbush, Ocean, and Park Avenues—designated specific "Jump Up Zones," where spectators could join bands to dance and party. The announcements also informed participants that no audio sound systems were allowed within one block of the parade route.

The "party" wound its way through the tightly policed and cordoned parkways, sponsored by multinational corporations and various Caribbean municipal and civic organizations of the city, in a public display of civic belonging. This was no flamboyant display of excess and hedonism. If anything, it was precisely the framing narrative for addressing the random fates of citizens like Louima. It requires twice the work for Louima and other immigrants as it would for a member of the affluent white majority to represent themselves as modern, civil, law-abiding citizens who are as concerned about welfare and safety as "the natives." To that end, the Brooklyn Carnival works as a vehicle of public engagement for diasporic identities from the Caribbean. Rather than a play on concealment and secrecy, the Brooklyn Carnival offers distinctly local forms of communal expression through neighborhood civic organizations, regulatory bodies formed by local communities, and group entities formed across broad alliances of immigrant and metropolitan interests. Thus it functions as a performance of the civic that activates new assertions of urban belonging.

Mermaids, Sharks, and the Sale of Dreamland

Under a powder-blue sky, on a dreamy, hazy, late June morning on Mermaid Avenue, in Brooklyn, strange watery beings wash ashore: red-haired Ariel, of Disney fame, in her resplendent red fish tail; splendidly yellow Neptune, in dreadlocks, on a mountain bike; a black-suited shark with a black briefcase, prancing menacingly, his black suitcase grimly displaying the phrase CONEY ISLAND SOLD. "Tranny sirens," sea creatures in fishtails of neon pink, purple, and shimmering blue, saturated fuchsias, magentas, and florid greens deliriously fill the sweeping vistas between Mermaid Avenue, Stillwell Avenue, Neptune Avenue, and Surf Avenue, in Coney Island. Punk rhythms and hard rock pound the macadam as mermaids in wheelchairs and stilts arrange the accessories of their finery in preparation for the start of what is popularly known as the Mermaid Parade: Brooklyn's alternative to the much-bigger annual Halloween Parade held in downtown Manhattan.

In Brooklyn, the aura is youthful, bourgeois, and funky, rather than hard, edgy, and uncompromising. Low-slung antique cars, motorized floats, and vans and trucks decorated with sea flora and mermaid paraphernalia gather their accoutrements of glitter, color, fishtails, and tridents amid naked torsos and alluring flesh. The dense, tightly packed crowd is very Coney Island, an agglomeration of utopianists, artists, lifestylists, environmentalists, activists, neighborhood regulars, and people from the city, out to catch a glimpse of the naked mermaids and watery creatures who have long been banished from the abandoned but fast-gentrifying waterfronts of Manhattan. Crowds of New Yorkers throng the street to celebrate the sea's importance to New York.

Begun as an artists' festival, by Dick Zigun, the founder of the Coney Island Artists Initiative, to celebrate the oceanic culture of New York City, the Mermaid Parade emerged in 1983 as a revival of the buoyant Coney Island Mardi Gras, which ran from 1903 to 1954. The Mermaid Parade occupies a special place among New York's many carnivals, as a place where art and dreams still comingle outside an economy of escalating real estate prices and the ever-urgent bottom line. On the first Saturday of every summer, an aura of the festive reclamation of Brooklyn's waterfront spirit permeates the seedy sea resort. The watery manifestations are intent on communicating the message to save Coney Island, to preserve its idiosyncratic, deeply nostalgic, and entirely New York quality of rough-and-tumble pleasure, still untouched by the harsh hand of suburban gentrification. A sense of crisis lurks in the parade, a deep melancholia at the passing of an era in which people consume Coney Island as an idea, a commodity, and a real place.

The Sale of Dreamland

On September 9, 2008, the gates of Astroland officially closed for the last time. Amid much protest and great public dismay, the seedy but greatly loved theme park of twenty-five years was bought out by Thor Equities. The sale captured an impending sense of ruthless change as New York's most idiosyncratic neighborhood, Coney Island, braced itself for more battle against dramatic incursion.

In 2005, talk of the billion-dollar glam-rock makeover of Coney Island by Joe Sitt, the owner of Thor Equities, rattled the city.[4] For many New Yorkers, Coney Island is a site of generations of memorable childhood moments, youthful escapades, and seaside experiences unique to New York. A Vegas-style makeover with a Disney-scale theme park is precisely what Coney Island is not. Sitt's projected plans for Coney Island included a resort paradise with retail and entertainment megaplexes, an indoor water park, a large four-star hotel, and a blimp that carries tourists over the area, advertising the hotel's name, THE BOARDWALK AT CONEY ISLAND, in giant font. Visions of such a gargantuan development's displacing existing neighborhoods, small businesses, and the area's distinctive cultural heritage generated massive public protest.

Initially Mayor Bloomberg cautiously supported the plans, much to the city's general dismay. Following vociferous neighborhood activism and protests, the projected plans of Thor Equities fell into disfavor with city planning. The City Planning Commission and the New York City Economic Development Corporation engaged in furious negotiations to purchase the area from Thor Equities, so as to rethink the proposed developments.[5] However, unwilling to pay more than the land is worth, the City Planning Commission was ultimately unable to purchase the land for the proposed sum of $105 million, which Joe Sitt turned down.[6]

The closing of Astroland echoed an earlier melancholia, the demise of other, more fabulous pleasure domes, such as the Elephant Hotel, a fantastical hotel built in 1886 in the shape of an Indian maharajah's palace. Other marvels of height and speed were Steeplechase Park, Luna Park, and Dreamland, all extravagant and garrulous palaces of fun from the early twentieth century, where New Yorkers could gather to enjoy delirious and weird entertainment at land's end.

Coney Island has traditionally embodied that aspect of New York City that is irreducibly New York. Now eerily poised to be reinvented as a Disneyfied theme park, it will more likely resemble Coney Island than be Coney Island.

Instead of the ramshackle beachfront hotdog, fries, and knish stands that defiantly remain amid the bleak scenario of bulldozed Coney Island real estate, with backdrops cryptically announcing "The Future of Coney Island," the future promises the scale of Las Vegas-style entertainment at Disney prices.

Instead of the derelict man who walks up to me with a catch in his throat to announce the impending closing of Astroland—"I don't know what I am going to do. I have worked here for twenty-five years"—I imagine a spruced-up, younger, shinier amusement park with faster carrousels and "better freaks," according to Joe Sitt.[7] The reinvention of Coney Island as a bigger, brighter, cleaner, Cirque du Soleil-style theme park reaches into the deepest, most resilient entrails of this hard city, to expose its historically shifting future.

CHAPTER 10

spirits of the necropolis, planes on the hudson

In *Toward Perpetual Peace*, Kant writes, "The Stranger has . . . a right to visit, to which all human beings have a claim, to present oneself to society by virtue of the rights of common possession of the surface of the earth." New York City stretches the limits of Kant's thesis on cosmopolitan belonging. Living in Manhattan is a bad addiction. In 2013, the city is unaffordable. All the interesting people—writers, artists, musicians, theater people—have moved to Brooklyn or Jersey City or Coney Island. Now everyone is going to Queens. My hipster students used to tell me that Williamsburg was the happening place. Now they say gentrified Bedford-Stuyvesant is the place to be. It used to be Dumbo, for a while: short for Down Under the Manhattan Bridge Overpass. One loses track of what's cool and hip once one lives in New York City. In a city of ever-changing landscapes, some sites continue to maintain their intensity through time. Christopher Street is such a place.

The coordinates of Christopher Street lend it a special topographical logic as a place through which to wander. While its contemporary aura vibrates with the hyped-up, jaded, seedy glamour of the overexposed yet curiously gentrified public park at the junction of a major transit hub, the New Jersey Path Train, it is one of the oldest pathways in Manhattan. On the surface, Christopher Street is the ur-place of modern desire, where all structures of modern love are exchanged and consumed. It offers the hard glitter of public sex and well-worn mores of street cruising, but its layered ambiance is that of an old street that has many transitions to tell of, not just the one rooted in the sexual revolution of the 1970s.

Its unusual width as a street and its broad base at the junction of Weehawken Street suggest the practical labors that demanded such a wide opening to the Hudson River in the early twentieth century—the transport of break bulk, noncontainerized freight. The earliest etchings of the island of Manhattan suggest a sandy little beach, a natural bay that formed out of the brooks running west from the wetlands of Minetta Brook, around today's Minetta Lane. These compelling early illustrations of a sandy cove right where today's Christopher Street Pier stands speak volumes to its history as a street of nomadic encounters, surreptitious trysts, and sex for money, where anything is available.

The historical records point out that when the British spotted this serene cove, they promptly built a pier and the first penitentiary on Manhattan Island, on the footprint between today's Washington and Greenwich Streets, on the north-south axis, and between Charles and Christopher Streets, on the east-west axis. This little tale of an idyllic cove turned prison complex in the early seventeenth century underpins what today is considered one of the most famous streets in New York and the mecca of the gay rights movement.

The pivotal role of Christopher Street in the history of New York City centers on the history of the Stonewall Inn, a bar where a skirmish between gay patrons and the police in 1969 triggered the most influential riot in the history of New York City, after the 1863 Draft Riots, creating a new political culture of cosmopolitan identity that is quintessentially New York: the lesbian, gay, bisexual, transgender (LGBT) identity.[1]

This claim to a new cosmopolitan identity centered on Christopher Street precedes its twentieth-century mooring, however. The location of Christopher Street as a port extends its history of transient identities and imagined reinventions back to the founding days of New York City's own struggle for metropolitan identity.

The annual Gay Pride Parade, in the last week of June, is always an exciting and elaborate display of cosmopolitan identifications. People come from all over the world to participate and consume what is essentially a utopian carnival of world becoming. The street is filled with people from out of town. Individuals of every disposition, hue, and belief system congregate along the pathways of the LGBT parade to generate a spectacle of heterogeneous love and possibility, deeply connected to a love for Christopher Street and the ambience it engenders as a place of hope for many who come here.

Desire fills the streets of the West Village at this time of the year. It is infectious and joyful, and offers a different imagining, even as the narrowing of the political arena on real issues creates despair—particularly the deep national denial regarding the needs of a growing population of LGBT-identified citizens. While the parade is largely experienced as an out-of-town event—many of the local residents around Christopher Street leave the city for the weekend in anticipation of the deluge of people—its euphoria permeates the air of the surrounding streets.

Every year extraordinary beings materialize on the sacred street in wondrous costume. The image still lingers on the street from a few years ago: three tall, celestial creatures on inline skates, with gigantic silver shimmering wings, floating through the streets of the West Village in balletic concerto. Another year, black Eros, covered in gold dust and gilt clothing, glided down Christopher Street, toward the opening lip of the street's end. This powerful image of love and Eros appeared at a time worn weary by war and loss, by guilt and complicity in terrible things unfolding on the world stage, whose aftershocks only registered as limbless, strapping youths riding the A train, some still in military fatigues. By contrast, Eros strolled from crowd to crowd, offering love and hope through touch and glance. The street shone in the harsh light of the afternoon sun and reminded the city that love remains a language of the urban. The street is the space of erotic possibility—real or fantasy—a chance apparition surprising one's dull existence into the pure, unplanned beauty of the passing moment.

The sighting of a black god with skin of gold dust on Christopher Street nudges the darker secrets of the street's past toward a compassionate present. Love forgives all, perhaps—but not easily. The reality of Christopher Street's history is one of a forsaken god, where freed black slaves owned the property between Christopher and West Fourth Streets, and between Bleecker and Spring Streets, during the early seventeenth century, under the Dutch. The eventual theft of black property under British colonial laws of slavery in New York systematically usurped this area, known as Little Africa, from free black ownership. During slavery, free black subjects were not entitled to the laws of primogeniture, whereby property could be handed down from generation to generation.

Now a black god fleshed in gold floats toward the widening berth of the Christopher Street pier as the sun's rays dramatically light up the ceremonial pathway. He is generous, playful, expansive, coy, and pure Christopher Street. Black Eros retraces a path of Greenwich Village's black history, a history that is painful and splendid in its embrace of public culture.

In the early seventeenth century, part of Christopher Street belonged to one Simon Congo. It is said that in the eighteenth century, if one stood at the foot of Christopher Street, one could see the shadowy forms of men hung in Washington Square Park.[2] Now the street is a converging point for youth of different ethnicities, seeking alternative forms of community, away from the realities of their own neighborhoods. Christopher Street continues to be a destination for black urban life. This pathway is a reworking of older black passages through black Manhattan, of which Christopher Street was a founding part, in the early history of the island. Here, ghostly seventeenth-century remains resurface through contemporary economies of desire.

Spirits of the Necropolis

Every Halloween since 1976, the Halloween Parade, started by Ralph Lee, on the west side of Greenwich Village, winds its way through the streets of downtown Manhattan. Lee initiated this big-puppet parade as a means of revitalizing the streets of the city and drawing children into the public spaces of celebration. The parade swiftly exploded into the gigantic event it is today. Spectacular creatures, beings, spirits, and demons haunt the Village at this time of All Souls Day, a Day of the Dead, every year.

The parade's route originates in SoHo, on Broome Street, south of Spring Street, on Sixth Avenue. Revelers gather in innovative costumes to march and frolic through what is largely a necropolis. Manhattan Island has many burial sites. Liberty Park, Washington Square Park, Bryant Park, and Madison Park are some of the better-known potter's fields, burial grounds for the city's indigents. Unbeknownst to many Halloween ghouls and goblins out haunting the streets of Manhattan, they materialize the spectral hauntings of many unseen ghosts of history, African American, Native American, and colonial. The area now traditionally associated with the Halloween Parade traverses three neighborhoods: SoHo, Greenwich Village, and trendy Chelsea. The underlying rationale for this route is the watery logic of the Minetta Brook, which connected Washington Square Park's swampy terrain to the Hudson River, across the footprint of Minetta Lane. This region was known in the eighteenth century as Little Africa. A plaque at the corner of Sixth Avenue and Bleecker Street commemorates this little-known piece of black history.

As caravans of otherworldly spirits waft uptown on Sixth Avenue, past formerly black-owned property on Bleecker Street and Christopher Street, the commemoration of ghostly unrest becomes a hedonistic and festive celebration of the palimpsestic mournings of a city of the dead. Police blockades during the parade arrest the swift flow of the city, converting it into a maze of closures and dead ends after sunset.

The whole of downtown Manhattan is largely a burial ground. There are an estimated twenty thousand people buried in Washington Square Park. Most bodies from this area have been posthumously disinterred and transferred to Randall and Ward Islands. The bodies of African American slaves buried north of Wall Street, extending all the way to the area around Central Park, renders the largely unmarked regions of black burial a public haunting for the city of Manhattan. Ghosts walk every day, reminding the city of its dense history of the lived and the dead.

This thick historicity comes alive at Halloween, as repressions are released and fears explode into tactile images of horror and the macabre. In the celebration in 2007, a decapitated man glided alongside burlesque characters with exposed genitalia. Sponge Bob walked with Larry Craig, beside an entourage of green Statues of Liberty. That year, many politicians were debating whether waterboarding was a form of torture, or just a coercive but legal method of interrogation. Former New York City mayor Rudolph Giuliani seemed to think the technique might be justified in extreme cases. The country embarrassingly violated international law as it condoned extreme rendition.

Suddenly the streets of the West Village, filled with grotesque and violent effigies, did not seem so fictive. Rather, they appeared to be a literal enactment of society's worst nightmares. Bleeding, mauled cadavers resembling George W. Bush, Dick Cheney, and Karl Rove grimaced and grinned out of Marc Jacobs's window on Bleecker Street. A hovering Donald Rumsfeld gleefully peered through the entourage. They cackled ghoulishly as revelers awash in blood sought a catharsis that was nowhere in sight, even as a half dozen Osama bin Ladens bobbed through the impossibly packed crowds of Halloween revelers. My own Viet Cong outfit seemed tame compared to the spectacularly brutal paraphernalia of war and pornography surrounding much of the mood around Christopher Street.

Planes on the Hudson River

Against the grey, windy afternoon light of the yellow-blue freeze, the bizarre spectacle of a plane floating swiftly along the icy-cold Hudson River, on January 15, 2009, derailed the West Side Highway. A double bird-strike had forced the pilot to head for the Hudson River as an engine exploded.

Partially submerged, traveling swiftly in the frigid river past the Water Park, on Jane Street, with a wing and its nose in the air, surrounded by ferryboats, tugboats, fireboats, and helicopters, the out-of-place plane reawakened every New Yorker's nightmare. We stared across the snowy landscape in disbelief—haunted once again by the unspeakable fear of planes falling out of the sky.

The plane sank swiftly, pausing and rotating as it is glided along the Hudson River current. Miraculously, the 140 passengers survived the downed Airbus Flight 320 from La Guardia to Charlotte, North Carolina; they were rescued in five minutes. But suddenly the Circle Line, the FDNY, and New Jersey and New York Waterways all pointed to New York City's dense local ecology of air and water circuits.

The Hudson River is designated as an emergency approach vector. On January 15, 2009, the river rose into the city's consciousness as a very busy intersection of transportation networks whose routes and regulations impact the city's neural health daily, without the public's full comprehension of its import on their lives (see map 5).

The airbus incident triggered thoughts of the peculiar junctures of plane crashes over New York City in the last decade: the World Trade Center planes; the crash off Far Rockaway, in Long Island, in 2002; the plane that flew into a building on the Upper East Side in 2006; and now this plane on the Hudson.

All water service in the New York area was suspended as the Hudson River event unfolded. This shutting off of water transportation, which connects many parts of New York City, brought back memories of the last time the ferry ways were stopped, during the World Trade Center collapse. Once again, the island's fragile ecology was brought to the fore. New Yorkers considered the implications of environmental contamination, and the resulting water safety and health implications, as the plane sank into the Hudson near Christopher Street. Rescue workers soon hoisted the wing of the sunken plane onto a tugboat and dragged the metal bird down to the Battery Park area.

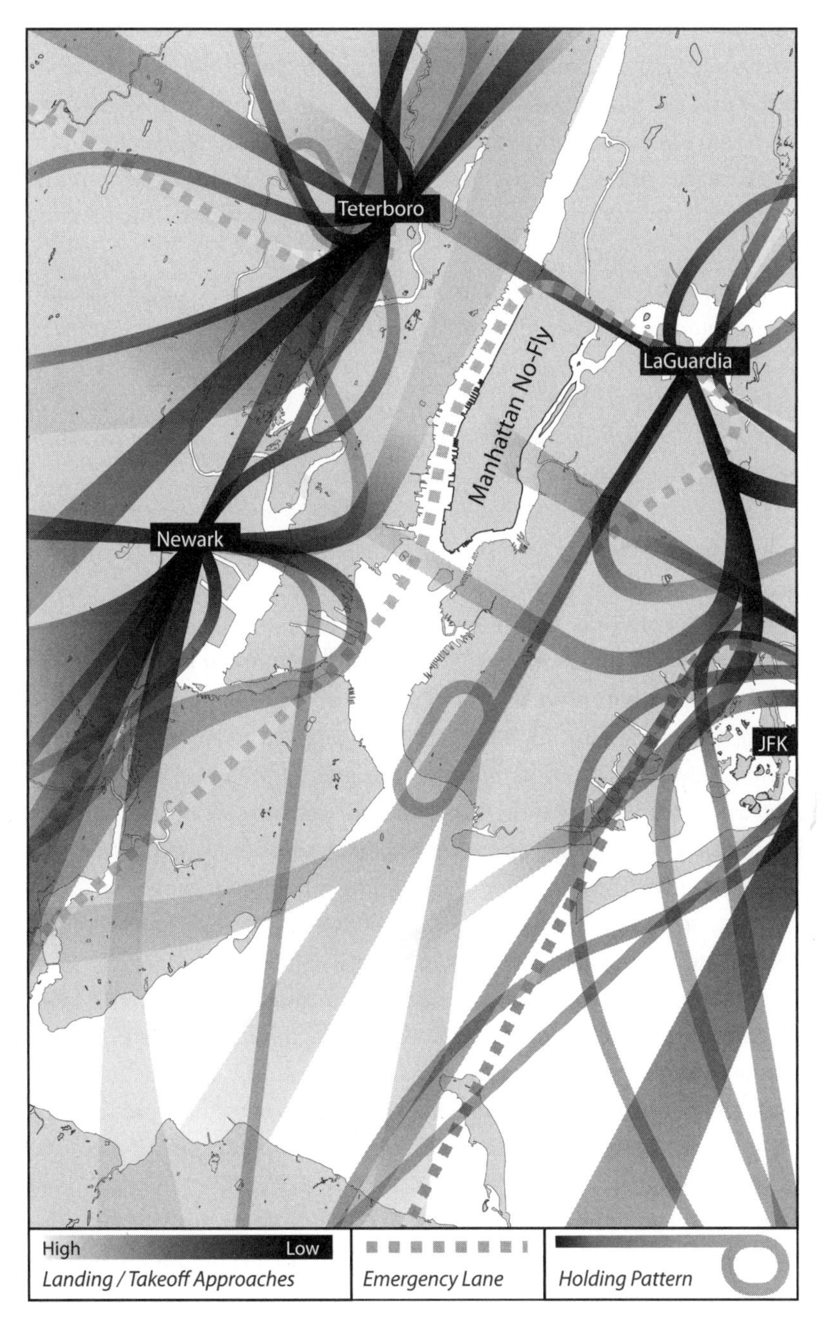

MAP 5 *Map of Jersey and New York City Airport Air Lanes and Waterways.* Courtesy of Daniel Hetteix.

The downed plane opened up the iced Hudson River to the city's view, reminding New Yorkers how tenuous the geography of the region is in its dense urban arrangements. The incident was a photo op for New York at its best, said Manhattan borough president Scott Stringer. But he combined exhilaration because of a day gone well with a familiar warning reminiscent of 9/11, to stay away from the piers, due to security issues surrounding the submerged plane.

Beyond the immediate drama of extraordinarily favorable circumstances and a catastrophe avoided, the incident once again accentuated the interconnectedness of shorelines. Survivors of Airbus 320 were taken to the New Jersey side of the Hudson River, as well as to the Manhattan side, creating a rare display of shorelines speaking to each other, two sides of a conversation that merge in critical moments.

Later, in April 2009, Air Force One swooped close to downtown Manhattan, heading toward New Jersey and startling people across the region. People in New York and New Jersey went through a momentary replay of 2001, fleeing high-rise offices downtown in panic. The incident turned out to be a poorly planned photo shoot for the presidential plane, which was conducted without notifying the citizenry of New York and New Jersey. The resulting mayhem highlighted once again the city's need to forge a better understanding of its uncertainties and its preparedness. A deeply concealed awareness that the city's strength is also its weakness, its interconnected geography being its vulnerability, erupts in moments of crisis, which have been many in 2009—including a bout with the H1N1 flu, in addition to the falling plane episodes—effecting a shift in how people on the ground experience daily life in the city.

New Yorkers are determined to ignore sensationalism of any kind, whether it is low-flying planes that spark raised anxiety or blue surgical masks in Manhattan subways during rush-hour traffic. Yet, the unspoken possibility of traumatic environmental calamity is never far from the daily commuter's consciousness. The studied disinterest, the expressionless dismissal of fear, is part of the civic expressivity that makes this city still an incredible place to live. People waited patiently underground for forty-five minutes in a packed subway train with no lights or air conditioning, in the height of summer, in 2008, and no one lost their cool. This calm defies easy explanation. It is a display of a deeper sense of connected reality in which everyone is dependent on everyone else's remaining calm in the face of a shared public life.

On August 8, 2009, another terrible calamity occurred on the Hudson

River. A helicopter and small airplane collided. After hours of uncertainty, it was clear that nine people had died. The Coast Guard, tugboats, the Army Corps of Engineers, the New York Police Department, and various security helicopters scoured the region during the first few hopeful hours. The initial projected toll was one dead. The uncertainty of where the other survivors might have floated raised the question once again of the interdependency between the New York and New Jersey shorelines. The West Side Highway was blocked downtown as emergency-relief vehicles combed the area for survivors. Once again, the delicate balance of shared airways and waterways, and the fluid zones of security and dependency between New York and New Jersey, foregrounded New Jersey as the silent sixth borough of New York City.

New Jersey: The Sixth Borough

The recent aviation disasters on the Hudson River and the environmental hazards raised by the two events reinforce the fact that New York City is a series of local realities. For the west side of Manhattan, the Jersey shore is a physical place, an environmental countersphere whose decisions and development impact New York critically. Whether the issue is industrial run-off, pollution, land rights, water rights, or, now, air space, the fate of New Jersey's coastline is linked to that of New York City's, and it is more urgent than ever that New Yorkers engage with this reality on a local basis.[3]

The width of the Hudson River belies the proximity of New Jersey's land mass to Manhattan and obfuscates the intensely trafficked air space that matrixes lines of potential conflict, if not disaster, on a daily basis. This peculiar ecological condition of two different governments coexisting in one cohesive environment is slowly transforming the way people in New York view their city and its future. In many ways, what happens in New Jersey is more likely to impact the quality of life across the Hudson River in New York City, in regard to growth, waterfront development, signage, air rights, visual field, and scale of development, than what happens in Queens or the outer boroughs of Brooklyn.

Falling Toxicity

The dramatic sight of planes on the Hudson highlights a subtle factor guiding the everyday probabilities of life in a vertical city: the potential threat of falling toxic or dangerous objects. The catastrophes of cranes falling on buildings, elevators falling from buildings, air conditioners falling from high rises, and planes falling from the sky, open up the tenuous logics governing life on the street.

In New York City, the concert of air traffic flows toward its three major airports—La Guardia, John F. Kennedy, and Newark Liberty—respecting a no-fly zone for larger aircraft around the Hudson River. Nonetheless, while residents of the city have little control over how federal aviation regulators monitor the skies of New York, the consequences of lax oversight have exposed New York City to another dimension of ecological vulnerability: technological safety as a transportation network issue, as opposed to a mere air-quality concern. In particular, the recent public awareness of flimsy regulations regarding low-flying small airplanes above Manhattan has focused attention on the fact that, along with land-bound technologies that impact human ecology in New York City, the matrix of air and water transportations also generates influential economies of risk around New York life.

The Staten Island Ferry is a popular object of such uncertainties. Accidents in 2009, 2010, and 2013 involving the crash of the Staten Island Ferry into piers in Staten Island are yet another demonstration of the interconnected networks of risk that structure a commuter's life in New York. In 2009 the ferry was carrying nearly 800 commuters, fifteen of whom were injured, when it collided with the pier. In a similar incident in 2003, a Staten Island Ferry carrying some 1,500 passengers slammed into a pier, killing eleven.

Such occasional fatalities expose the inherent vulnerabilities of living in a dense, multispatial transportation ecology of air, water, roadway, and underground travel options. The potential for planes to fall, buildings to crumble, boats to collide, bridges to collapse, cars to crash, and subways to become stranded underground, generates an awareness of "falling toxicities" that are endemic to city life in New York. However, this is not a premonition of catastrophe, but rather an opportunity to examine ecological sustainability. How can a metropolitan region such as New York integrate its complex air, water, road, and rail transportation networks in ways that attend to New York's future growth without sacrificing its citizens' rights to safer skies and less stressful commutes? This is a question with which New Yorkers must be increasingly engaged.

CHAPTER 11

governors island | *maritime pasts, ecological futures*

At the tip of Manhattan, there is a small landmass called Governors Island. Standing on the southern point of the tiny island, one is gripped by the intensified drama of the confluence of bodies of water. Here, the Hudson River, the East River, the Verrazano Narrows, and New York Harbor swirl at the island's tip, creating the Upper New York Bay. Ahead, through the narrows in the distance, the waters surge out onto the Atlantic (see figure 11.1).

Before January 2003, the mysterious island had remained intractable, a forgotten aspect of New York City—inaccessible to the general public, gently contoured against the Atlantic haze, a bearer of early New York history whose connection to Manhattan, though geographically intimate, had been formally separated through much of the city's urban past. Looking across at Governors Island from Manhattan's waterfront, one recalls the observation by the French philosopher Emmanuel Levinas, "These cities (of refuge) are to be made neither into small forts nor large walled cities . . . they are to be established only in the vicinity of a water supply and where there is no water at hand it is to be brought thither . . . and if the residents (of any one place) have fallen off, others are brought thither, priests, Levites and Israelites."[1]

One hundred seventy-two acres and barely eight hundred yards off the southern tip of Manhattan, Governors Island was a beautiful, ghostly presence, home to a prisonlike structure forbidden to the public for all its

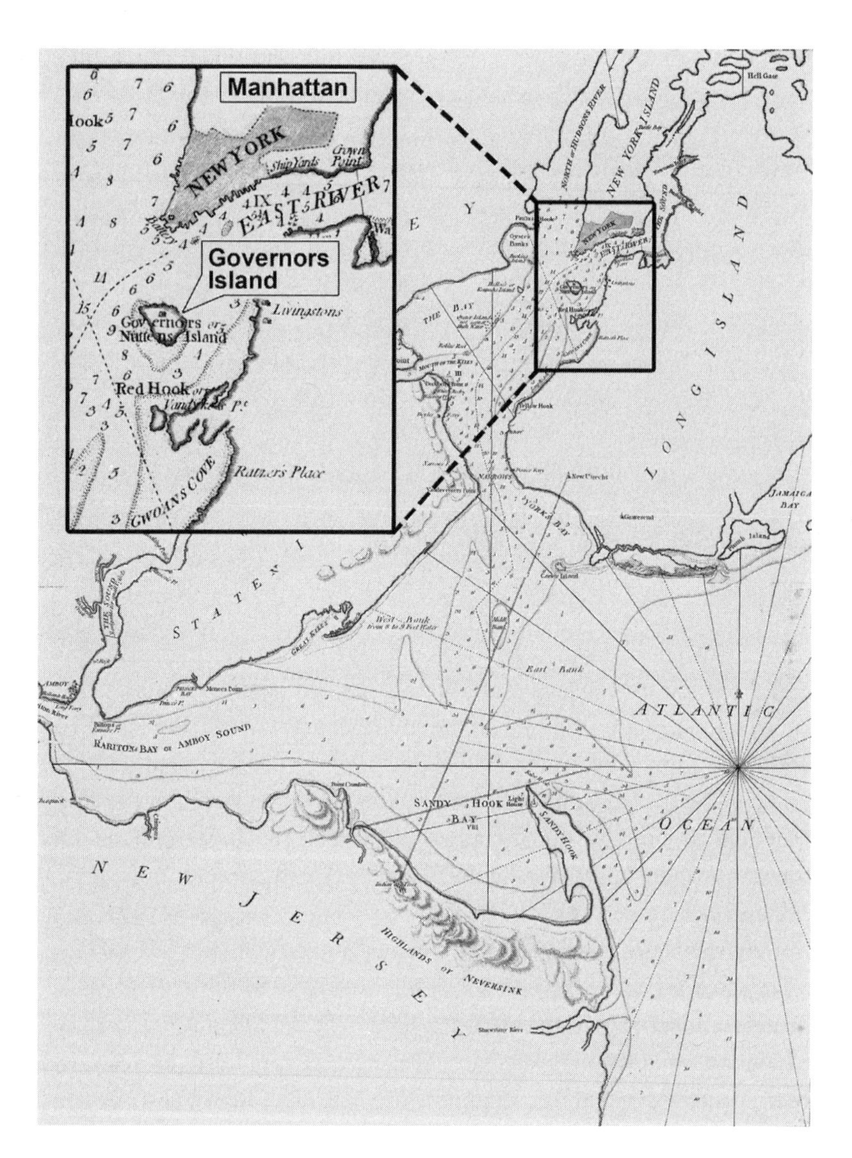

FIGURE 11.1 *Map of Noten Eylant, 1776: Chart of the Entrance of Hudson's River, from Sandy Hook to New York*. Courtesy of the New York Public Library.

colonial and modern history, barring a brief moment under Dutch rule, from 1652 to 1664, when Noten Eylant was made available for the public's leisure.[2] Over its four-hundred-year history, the island came to be a fort, a prison during the Civil War, a U.S. Navy base, and a Coast Guard base.[3] Governors Island, formerly a U.S. military base, was subsequently shut down during the 1960s, and its title given to the United States. In 2003, the City of New York bought Governors Island for one dollar. Its ownership from then was shared by city and state. Recent talk of single ownership of the island by New York City is shaping plans for future park development on the island.

Originally a Native American settlement for over a thousand years, the Mannahattas called the site Pagganack Island.[4] Its colonial Dutch name was Noten Eylant, or Nutten Island.[5] The precolonial landmass, before landfill, was tiny. The Dutch bought the island for the price of two axe heads, some white beads, and a handful of nails, according to an embossed plaque mounted on the island in 1951 by the Government of the Netherlands. A little-known fact about Noten Island is that it is the site of the earliest Dutch settlement in New York City, established before the sale of Mannahatta to the Dutch. The first group of Dutch settlers stayed on Noten Island till the Dutch Fort Amsterdam was constructed on Mannahatta. Archaeological remnants of a wind-powered sawmill built by the Dutch West India Company in 1625–26 are the oldest Dutch remnants found around New York City to date.[6]

Governors Island serves as a reminder of the changing technologies through which New York City's prominence as a world city has emerged. The structures of forts and battery on the island's interior invoke the invisible history of New York's prominence. During the eighteenth century and the nineteenth, hidden amid fortified barracks and dungeons, the island concealed the military armaments undergirding New York's free and democratic demeanor. The little island off Manhattan's tip was the residence of three armed forts equipped to defend New York: Fort Jay, Castle Williams, and the South Battery. The presence of a British cannon on the island, dating to the Revolutionary War, is the sole remaining marker from the British occupation of Governors Island from 1691 to 1783. A pen-and-ink illustration from the Revolutionary War period, of a man on a peculiar contraption, half-submerged in water, suggests the militarized nature of the island's waterfront. Halfway between a rudimentary standing bicycle and a suspended chair, the contraption, known as "the Turtle," was an early prototype of the submarine. Efforts to develop torpedo techniques during the Revolutionary War, while unsuccessful, paved the way for the island's emphasis on maritime warfare technology up to the First and Second World Wars.

Multiple logics shape this distinctive island. The fort city is the most striking planning logic that guides the eye and mobility; that logic still structures how people experience the island today. Cars are not permitted on the island. Limited ferry access to the island, between June and October, on weekends only, coupled with limited transportation options, walking and bicycling, along with pedicabs and tandem bikes, creates a sharp disjuncture between the military feel of the landscape and its current use as a public park. Car-scale roads from prior eras in which military vehicles were used make the enforced pedestrian-scale perambulation a dislocating and surreal experience for the visitor.

Governors Island's fort-city logic is a modern carceral-scape. It creates the unique juxtaposition of sinister penitentiary violence and breathtaking harbor vistas of the most spectacular skyline in the world, downtown Manhattan and the Brooklyn and Manhattan Bridges, with the New Jersey shoreline and Ellis Island across the waterway.

Walking through the eighteenth-century stone ramparts of the forbidding Fort Jay, a star-shaped construction built in 1794 with a drawbridge and a dry moat, the modern visitor enters colliding spatial frameworks. Quotidian reminiscences of daily life emerge within the militaristic logic of the fort city, as neat kitchens with modern amenities sport lace curtains amid the gunnery on the battery. The fort was built to contain barracks for a thousand men. It housed a well for water, a hospital, and kitchens.[7] In its later incarnations, it became a slice of suburbia, with a Burger King outlet, theater, bowling alley, and parking lots, creating a provincial feel at the entrance to the most dynamic city in the world. A row of ten-inch Rodman guns still poised across the battery of Fort Jay foregrounds the heavy military arsenal protecting Manhattan's harbor over the last two centuries.[8]

The historic import of this well-preserved fort city is far reaching. Governors Island remains a material marker of the shifts of surveillance and warfare technologies from land- and sea-based informational systems to the modern air and electronic-data systems that rendered obsolete the gigantic fort-port cities that were constructed as signs of military prowess at strategic vantage points, such as islands, cliffs, mountains, and shorelines, along the Dutch East India Company's, and later the Dutch West India Company's, outposts.[9]

The restricted access to the island has preserved the layered contexts that have transformed Governors Island through time. Dutch, British, Native American, and American influences all lie in the thick sedimentation of Governors Island's soil, as has been revealed in recent archaeological digs, where Native American settlement debris of four thousand years ago has surfaced alongside Dutch remains from the early 1600s.[10]

Penitentiary Aesthetics

Governors Island is an archive of disciplinary logics. Surveillance imperatives drive the design of buildings, landscaping, and the island's traffic flows. Its structuring sensations, as one cycles around the island, evoke a dramatic contrast between leisurely freedoms and military surveillance. Drawbridges, moats, guns on ramparts, cannons, and a circular sandstone battery complex that doubled as a prison, with extensive fortifications, reinforces the penitentiary aesthetics of the island. The remarkably unaltered state of the barracks and the round, cheese-shaped Castle Williams, with its three floors exposing prisoners' cells, generates a vernacular of force and regimentation that is oddly transformed into a theme park as tourists cycle through these grim carceral constructs.

Castle Williams was a notorious prison for Confederate prisoners of war, who were held in its dank cells without access to the outside. Built in 1811, Castle Williams contained as many as 1,500 prisoners.[11] The circularity of the battery on Castle Williams, along with the circular shape of Castle Clinton, at the tip of Manhattan, which was completed the same year, leave a visual vernacular of New York's colonial history. Together they open up less-visible questions of power, repression, violence, acts of war, and territorial occupation that underlie such architectural statements.

The forbidding façade reminds us of the brutality of the Dutch and British occupation of Native American land, without being accountable to that period of history, since the structure dates to only 1811. Furthermore, Castle Williams asks the more challenging question as to how the city of New York should develop a site so rich with New York's military history. Should it coat the rawness of the island's brutal edifices with more contemporary details or leave the island's fortifications as markers of the island's journey to a place of public enjoyment?

Confronting the penitentiary logics of Governors Island is to enter a shadowy world that shaped the prominence and security of New York Harbor as a world destination. Governors Island signifies the culture of security and fear of conquest that characterized the founding of the city of New

York, marked by the departure of the British regiments from Governors Island, in 1783. While the military role of Governors Island diminished after the Second World War, the island remained a strategic location for aviation during the 1930s. Fiorello La Guardia, the mayor at the time, campaigned considerably to make Governors Island a public airport.

Island of Dreams

In 2007, the debate as to what to do with Governors Island became a public issue, as Donald Trump made a bid to buy it and privatize the property, to turn it into an island of luxury condos. The resulting public outcry and counterproposals by the city resulted in a competition to propose a design for the island. A Dutch architectural firm, West 8, won the competition. Its vision for a unique public park with bike-sharing arrangements and a system of aerial gondolas connecting Brooklyn and Manhattan, designed by Santiago Calatrava, has been received with great interest by the public. However, other competing development interests with mixed impact implications for the island's sustainability have also been floated, such as a New York University campus with student and faculty housing, a plan that is a terrible idea for the general public at large; a replica of the Globe Theater, to be constructed within Castle Williams's structure; and a high school, the New York Harbor School. As of 2011, the island's future still hangs in the balance, as the city's fiscal crisis affects potential development plans around New York City.

In the meantime, Governors Island offers many opportunities for public enjoyment. It has emerged as a destination for New Yorkers of all stripes. Picnic island, pleasure island, party island, art island, performance island, and leisure island are some of its emerging identities. Artists of all persuasions have been welcomed to present their work through festivals, performances, and art exhibitions on the island.

The challenges of organizing outdoor events are many. One of the quintessential obstacles to performing outdoors in New York City is the prerequisite to acquire a permit, should twenty or more people gather in a public park. In addition, in order to take any equipment onto Governors Island, one needs a permit or a sponsor. Nonetheless, the site offers an attractive counter to Manhattan, as it is a highly restricted environment with no residential or commercial enterprises. Despite the red tape involved, the utopian space of Governors Island allows a sensation of controlled freedom. Artists can create public sculptures and installations on macadam and in buildings within the parameters of the Parks Department rules.

The Governors Island Ferry is a designated ferry route with hourly service that, until recently, ended by sundown. More recently, the ferry service has been extended until midnight, making Governors Island a party destination for night revelers. This idea of an island of revelers is a whole new concept for a public park, a concept not realized on such a scale before. Governors Island's geography allows it to entertain utopian experiments. So far it has successfully experimented with a summer festival of public art, international public installations, masquerade parties in turn-of-the-century clothing, an annual polo tournament, environmental fairs, an ecological educational site, and a faux nouveau Dutch island.

During the first two weeks of September 2009, Dutch curators and artists initiated the New Island Festival, a two-week art and performance festival on Governors Island. While acknowledging an underlying claim of a historical past that has since been erased, the festival was a celebration of prominent Dutch performers transplanted from the annual Oerol festival, on the island of Terschelling, Holland.

During this two-week period of Dutch extravaganza, the whole of New York was inundated with many things Dutch, such as Vermeer's famed "Milkmaid" painting visiting the Metropolitan Museum of Art, Henry Hudson's *Halve Maen* sailing around Manhattan and the New York harbor, and Johannes Vingboons's rare maps' being exhibited at the South Street Seaport Maritime Museum.

The prominence of Governors Island as a major New York destination emerged in New York's public consciousness during the summer of 2009, aided by the marketing and promoting of the island through the news media, word of mouth, and the buzz of its destination as a center for international art and theater.

Governors Island's transition from a secluded island in the middle of one of the world's most prominent harbors to a major public destination is a reminder of how New York's landscape is constantly shifting in its inflections. For many seasoned New Yorkers who claim to know everything about New York, the views from Governors Island are a completely new experience. One realizes that the twin, circular batteries of Castle Williams and Castle Clinton, both built in 1811, are symbols of a New York that once had an architecture of fear as its most prominent edifices. Today, these two remnants from an earlier period of colonial encounters are reminders of older narratives of occupation and conquest, of arrival and departure, that shaped New York's image of itself.

Watching the replica of the *Halve Maen* float around New York Harbor in September 2009, one saw the historical importance of Governors Island, the site of the first colonial settlement following Henry Hudson's "discovery" of New Amsterdam. The island appears once again ready to reinvent itself as a destination where the city's beginnings and its future visual imaginings engage each other.

Henry Hudson and the Halve Maen

An eighty-five-foot wooden ship with bright blue and yellow trim sailed up the Hudson River on September 1, 2009. It flew ghostly sixteenth-century symbols: the foremast and mainmast each bore a United Provinces flag; the jackstaff flew the flag of the Dutch East India Company (VOC, its Dutch abbreviation). The majestic sight of six sails on three masts was surreal to behold, because it was a replica of Henry Hudson's *Halve Maen*.

A flotilla of Dutch barges with sails of blood brown and manuscript yellow followed the archaic ship. The sails of some of these boats were also embossed with the VOC insignia, the maritime sun image the company often used on their cartographic documents. The scene was particularly strange and disconcerting for this viewer, who hails from a former VOC colony, and for whom the idea that a colonial occupier would float around in period costume glorifying its colonial legacies is bizarre—though in this case spectacular.[12]

The *Halve Maen*'s appearance, in September 2009, inaugurated the quadricentennial celebrations of Henry Hudson's arrival in New York Harbor, in 1609. Hudson, an employee of the Dutch East India Company, came upon the great harbor of the North River, as the Hudson River was first called, and the shores of Mannahatta, during his third voyage in his search of the western sea passage to Asia. Hudson himself vanished on his fourth voyage, set afloat on Hudson Bay in a little boat with a handful of men and his young son, during a mutiny aboard his ship.[13]

The scale of activities around the quadricentennial included commemorative celebrations in communities in upstate New York: in Ulster County, the Hudson River Valley, Woodstock, Essex County, Cold Spring, Athens, Peekskill, Nyack, and Haverstraw. Exhibitions exploring the influence of Dutch food, Dutch finance, Dutch horticulture, and Dutch cartography were among the many discussions unfolding across New York State. Documents from Amsterdam on display at the Museum of American Finance, in New York, in September 2009, elaborated on the financial cultures shared by New Amsterdam and Amsterdam during the first two hundred years of colonization. This included the oldest known share certificate, prepared by the Dutch East India Company in 1606. On Staten Island, a group called Staten Island OutLOUD conducted a participatory reading from the journals of Robet Juet, first mate on the *Halve Maen*.

Set against the Brooklyn Bridge, a stunning view of the *Halve Maen* sailing gracefully before the skyline of Manhattan greets the eye. It is a breathtaking sight from Governors Island, looking north to the skylines of Manhattan, Brooklyn, and New Jersey. The harbor is a dreamscape of white, brown, yellow, and cream sails against the metal and brickwork of the Manhattan skyline, Brooklyn's burnt-red-brick palette, and New Jersey's metallic grey sheen.

"Want to know what I am thinking?" says a sprightly older gentleman standing next to me. "I am thinking—these are vistas never seen before. I am thinking that in all my seventy-three years, there hasn't been such a sight to see. What incredible views! These are new vistas, unprecedented perspectives."

The original *Halve Maen* was viewed from marshy banks, forests, swamp, and now-flattened hills. Framed by the choppy, gray-green ocean waters of the Verrazano Narrows and the New York waterfront, the seafaring spectacle in 2009 is indeed a perfect new vista.

From September 1 through September 13, 2009, the Hudson River and its environs were home to a festive panoply of ships sailing around Manhattan's waters. On September 8, 2009, flat-bottomed Dutch barges sailed downriver early in the morning, a memorable sight of brown and white sails bobbing down the Hudson River, barge after barge. The scene is unlike any seen in New York's recent history, the city's waterways filled with every possible seafaring vessel in a full embrace of the city's waterfront.

Culminating the festivities on September 13, 2009, a commemorative entourage of NATO warships, Dutch barges, ferries and tugboats, sailboats, sloops, kayaks, water taxis, and speedboats coursed through New York's waterways, all gracefully escorting the colorful replica of the *Halve Maen* up and down the East and Hudson Rivers, initiating the first Harbor Day in New York City's history.

The spectacular staging of New York City's operatic seafaring mise-en-scène emphasized the archipelagic structure of New York City. It was a rare month when one experienced the sensation of living in a city by the sea.

Boats sailing up and down rivers, dramatic sails sighted at odd times of the day and night, maritime paraphernalia and regalia displayed in museums around the city all contributed to generate the distinct maritime feel of a city structured by water—a city made up of many islands, surrounded by water, connected artificially by an idea of metropolitan becoming that has transformed these myriad little islands into a cellular unit called New York City.

Largely organized by the Henry Hudson 400 Foundation as part of the NY400 festivities, the four hundredth anniversary of Henry Hudson's voyage invited a widespread cultural revisiting of Dutch New York. The website of Explore NY400 offered twenty pages of programming for the month of September 2009, ranging from curated exhibits on the roots of Dutch New York and New Netherland history in the Hudson Valley region to reenactments of Henry Hudson's voyage, involving high school students as crew aboard the replica *Halve Maen*.[14]

Celebrating a colonial encounter is a fraught undertaking. In settler colonies such as the United States, the narrative is one of cultural reclamation for the colonizing force that relinquished power. For modern viewers of such volatile encounters, the relationship of the historic document to the possible meanings it offers is frequently contradictory, both enlightening and revealing of historic gaps and silences.

In particular, the hand-painted maps of the Dutch East India Company and its prolific cartographer Johannes Vingboons, on view at the South Street Seaport Museum, were sweeping testaments to the breadth and scope of colonial Dutch involvements on a global scale. However, the maps in the South Street Seaport exhibit also demonstrated that the colonization of Mannahatta occurred in conversation with other colonial sites, such as Batavia, Surinam, Puerto Rico, Havana, Recife, Cape Town, and Cochin. These rare documents allowed a glimpse into possible interactions unfolding between colonizer and colonized among different Dutch colonial sites.

Questions arise amid the careful Dutch documentation of land acquired from the Native Americans: What did the Native Americans think of the strange people living on Noten Island, and later in New Amsterdam? Did the Dutch colonies of Pernambuco and Curacao impact New Amsterdam's governance? They shared the trafficking of human beings to New Amsterdam, so surely there must have been more than just human bondage being circulated between these Atlantic Dutch ports. One wonders about the silenced voices of indigenous peoples whose lands emerge in these gilded maps as trophies of the Dutch empire. These gaps of history hidden in the

Dutch archives are an important dimension to this emerging story of the Dutch seventeenth-century in translation.

Celebrations on the scale of the quadricentennial always risk reductionism — particularly when the site of the elaborate return to the colonial past is today the center of the financial world, whose relationship to the colonizing presence being celebrated was one of utilitarian desire. Amid the panoply of Dutch reclamations of all aspects of New York's life, an oversimplification of the complexity of Dutch New York was unavoidably unfolding around New York City. The mysterious story of the British expeditionist who worked for a Dutch corporation and introduced Mannahatta to the European imaginary fades against the plethora of pure Dutch kitsch, uninflected by its extensive seventeenth-century global history of conquest and encounter. Instead of the extraordinary worldviews garnered by early Dutch travelers of the Atlantic, Pacific, and Indian Ocean worlds, of which Mannahatta was just one outpost, the image grasped by the passerby is of a provincial Dutch village transported, without the ravages of encounter and syncretization, to the little hamlet of Bowling Green Park, the originary site where the first Dutch settlers set up Fort Amsterdam on Mannahatta island.

Yes, a faux Dutch village on Bowling Green Park, replete with a white windmill rotating in the wind, marks the symbolic four hundredth anniversary of the arrival of Dutch settlers in Manhattan, in 1609. The scene is one of unapologetic down-market Dutch kitsch, without any attempt to cite the Dutch East India Company's extensive colonial baggage. Sales of Dutch clogs, tulips, souvenir windmills, and Gouda cheese reduce the historic encounter to a clichéd representation of seventeenth-century Dutch culture, along with the cheap Vermeer reproduction propped against a booth — as if to say, "See? We are really as Dutch as the Dutch masters portray us." One is reminded more of Solvang Village, the faux Danish village in California, than the greatness of the Dutch maritime empire of the sixteenth century that is encountering various continents of indigenous societies.

In tandem with the heritage reproductions on Bowling Green Park, Dutch performers on Governors Island offered a whole ten days of Dutch culture in 2009. New York's summer crowds were invited to immerse themselves in avant-garde Dutch art and performance at the site of what was, four hundred years ago, an early Dutch settlement purchased from Native American peoples. It was a thrilling event, filled with unexplored encounters with the landscape of Governors Island, as Dutch performers used the island's unexposed landscape for nighttime environmental pieces. Previously prohibited areas were opened to the public for the first time during the New Island Festival, giving it an allure distinctive from that of other art festivals. The island was made accessible and strange to the public of New York City, as it allowed audiences to remain on the island after dark for the first time. One could not help imagining the company of ghosts from many eras of history—the Native Americans, the Dutch, the British, the Confederates, and the Americans—as one wandered the island's penitentiary spaces at night.

The Dutch New Island Festival forced a new open-spiritedness onto a forbidding island. The festival pamphlet read: "As Henry Hudson did 400 years ago, a new wave of Dutch pioneers will be journeying across the Atlantic this September to a small island in the middle of New York harbor. From their settlement on Governors Island, this collection of artists, dancers, musicians, singers, and cooks are here to re-capture New York."[15]

Tongue-in-cheek colonial references notwithstanding, the cornucopia of avant-garde creativity from the Netherlands, inundating Governors Island, was a delightful gift to the city. Still, the extravaganza invoked the event of Dutch colonialism without addressing the historic encounter on Noten Island between Dutch settlers and Native Americans before the establishment of New Amsterdam.

Native American settlements had occupied Noten Island for over a thousand years before the arrival of the Dutch, as shown by the archaeological debris unearthed on the island.[16] Hence the uninflected celebration of Dutch colonial culture as a relationship between Amsterdam and New Amsterdam, rather than as a more interdependent association between Native Americans and the early Dutch, as well as between New Amsterdam and its regional ports of Curacao, Pernambuco, and Puerto Rico, was misleading. One left the festival with a peculiar iteration of modern Dutch nationalism expressed as a provincial claim from a remote past, rather than as the celebration of a seventeenth-century Dutch internationalism, which was the Dutch East India Company's most remarkable achievement as a major global corporation with outposts on every continent. Still, a lesson about abandoned islands' being turned into havens for art, performance, and music was certainly one to be absorbed from the Dutch.

Maritime Pasts, Ecological Futures

On Saturday, May 22, 2009, at nine-thirty a.m., cirrus clouds and choppy green waters greeted the Christopher Street Pier crowds and intrepid swimmers who had lined up to swim the Hudson River. It was Swim New

York's Hudson River Swim. The pier was swarming with black-wetsuit-clad, barefoot bodies. At ten-thirty the flock of orange swim caps lined up at the Water Taxi jetty and entered the sixty-two-degree Hudson River, one wetsuit at a time—about a hundred in all. They huddled in clusters of about thirty and broke off into a rapid swim amid tugboats, sailboats, kayaks, orange buoys, incoming shipping traffic, motorized river traffic, and the jutting piers of the island.

The sight was thrilling: shoals of humans swimming southward like eels toward the mouth of the narrows. They fanned out like fish and traveled downriver swiftly, finishing at the World Financial Center jetty, at Battery Park, before the jogging spectators could get there. Just another New York day marking the boundaries of how we imagine and desire the city. Finally, the island's waterfront was reclaimed as clean enough to swim.

Opportunities for New Yorkers to swim around their city by participating in different swim events have broadened over the last decade. The improbable sight of human biopower amid the thunder and volatility of industrial traffic and reclaimed piers is a reminder of the delicate river ecology of the Hudson River. Swimming reclaims the waterline in a physical sense. New York's maritime pasts are merged with its idealized metropolitan future as a habitable waterfront fit to swim in—still a dream, given the erosion of decades-old protections along the Hudson River.

The act of swimming in New York's rivers and around Manhattan in a counterclockwise circuit functions as a mnemonic device for the Clean Water Act of 1979. It reminds New York that its rivers need to be kept safe enough to swim in, that the hard won protections of the Clean Water Act are easily corroded. Spectators watch with considerable unease and wonder as hardy swimmers take on the city's choppiest terrain at the tip of Manhattan's landmass. On a good day, detritus of all sorts floats by the piers: plastic bags, wrappers, soda cans, plastic gloves. Is the water clean enough for human contact? It is a question forced upon the island of Manhattan, as it is reclaimed momentarily in all its poetic possibility.

Swimmers traversing the western edge of the island's Hudson River Park, heading to Battery Park, invoke the history of landfill and utopian land reclamation. What was largely a terminal for transporting toxic waste from the World Trade Center site on barges after 9/11 is now the scene of sport and pleasure. Against the backdrop of the bay with the Statue of Liberty and Ellis Island, the city's struggles seem to recede into a distant past. Sheer endurance reigns over the pressures of brownfields, superfund sites, and the marshlands of New Jersey. The intimacy of human life accentuates the abstraction of the struggle for clean air and clean water into a concrete moment.

In July 2011, a fire in a sewage treatment plant in North Manhattan resulted in a catastrophic environmental disaster. Millions of gallons of untreated wastewater spewed into the Hudson River, contaminating its vicinity and surrounding environs. The River that Flows Both Ways, as the Hudson River was known to the Lenape peoples, was transformed from a nurturing water source that was an environmental success story in the 1970s to a flow of contamination affecting large areas around New York City. Customarily a source of health, respite, and nourishment across the length of its trajectory, the Hudson River emblematizes contamination once again. Its image as a biohazard once again brings to public attention the vulnerability and interdependency of local ecologies.

The sewage spill in the Hudson River threatened the precarious balance between urban infrastructures and environmental concerns. It reasserted the need for environmental vigilance among cities along the Hudson River affected by the extensive seepage of sewage upstream and to coastal areas further down. Over the years, cities along the Hudson River had grown used to the benefits from the battles leading to the Clean Water Act. Enjoying the majestic Hudson River as a place of peace, enjoyment and leisure had become a way of life. Swimming in the Hudson River around New York City and further up the river became a popular way to claim the river's rehabilitation as a safe place. What was obscured by the pastoral picture of the Hudson River's success story was the need for greater environmental oversight across cities, to maintain their shared rivers and water sources.

The issue of daily sewage seepage in the Hudson River during overflows and storms is one such matter that has continued to command the attention of the city. Keeping the river clean involves a continuous commitment to infrastructural maintenance, including the overhaul of aging sewers in New York and other cities along the Hudson River. This requires fiscal and planning investments that are frequently relegated to the back burner because they are not high-visibility projects that bring public acclaim to city governance, till an accident such as the fire at the North River Wastewater Treatment Plant.

The wastewater spillage in 2011 evoked a regional sense of shared risk and responsibility. Cities are increasingly demanding public environmental accountability from their governance structures. Along the Hudson River, responses from beaches and public parks during the sewage spill displayed considerable frustration toward, and a higher expectation of delivery from, city governance on the matter of hydrological sustainability. People want safer waters, safer rivers, safer cities. They expect their cities to ensure the durability of what the political theorist David Held identifies as nonsubstitutable resources, such as the Hudson River.[17]

New York's bodies of water include watersheds, estuaries, and rivers. New York Harbor, also known as the Upper New York Bay, is fed by the Hudson River and the Gowanus Canal. The harbor also connects to the Long Island Sound through the East River, a tidal strait. Another tidal strait, the Harlem River, adds to the well-ventilated, navigational circularity around the island of Manhattan. The global implications of New York's maritime pasts link the port to its colonial cartography, the trading posts of the Dutch East India Company that eventually became posts belonging to the Dutch West India Company. This fact was physically anchored in the recent discovery of a Dutch East India Company cannon, unearthed out of the rubble of the World Trade Center site.[18] The bulky iron artifact's material presence cements the city's cultural link to its sixteenth-century past. Its sheer presence demands a fuller understanding of New York's weight as a port of entry when viewed on a historical scale.

From the vantage point of the city's long history, many new immigrants to New York are products of the Dutch East India Company's colonial history. Australia was known as the New Netherlands, and Formosa (now Taiwan) was considered one of the prized Dutch outposts, next to Batavia and Malacca in the extended map of outposts belonging to the Dutch East India Company. Sitting on the E train, then, observing the heterogeneous demographics of Queens, where many new immigrants live, one sees old cartographies and new maps. Modern nations have transformed, their new identities rooted in older connectivities and linkages now forgotten. History is circular, and the history of New York City even more so, with its ebbs and flows, eddies and swells.

CHAPTER 12

after hurricane sandy

Water in New York has a special significance. An archipelagic city, New York is surrounded by water, yet because of its density and skyline, people rarely get to see water lapping at the shores of their city. If you take a walk along the newly created greenways skirting the city today, you will see gentle waves splash the sides of the city, depending on the tide and the wake of the ships sailing by. Water is for New York a source of respite, a restorative, a precious commodity whose urban history involves a steady disappearance from its contemporary landscape, as noted by Colonel Viele in his engineering map of 1865 (see figure 1.1). Coney Island became a focus of this fascination for the elusive element of water at the heart of metropolitan living.

On October 29, 2012, the city's relationship to water changed irrevocably. A new dread has crept into New York's unconscious. It is a historically unprecedented unease with what the sea portends for New York's extensive waterfront communities. Water is no longer an innocent and leisurely commodity upon which the business of the city is transacted. It is now an ominous threat. Where the ocean used to beckon, at the Verrazano Narrows, at Far Rockaway, at Breezy Point, at Staten Island, at Sandy Hook, the shoreline evokes a scene of homes destroyed, a way of life washed away unexpectedly.[1]

What is shocking about this scenario is not the fact that a storm of unimaginable magnitude wreaked so much devastation along New York

City's waterfront, though the fury and enormity of the storm was unprecedented in its scale and force. The shock comes from the extent to which New York has been unprepared for its ecological context as an archipelagic metropolis by the sea.

Storm Surge City

New York is a city vulnerable to storm surges. Climatologists have been warning for years that storms of considerable magnitude are inevitable for the city. Out of 140 port cities in the world, New York stands fifth for risk of flooding from storm surges.[2] Tropical Storm Irene demonstrated, in August 2011, New York's precariousness, when it stranded without electricity half a million consumers in the region.[3] Much of the city's transit networks, such as its subway system, as well as its water tunnels and electrical conduits, lie fifteen stories below sea level.[4] Planning infrastructure to prevent catastrophe is expensive, but cost effective in the long term, if the city is going to survive the reality of rising oceans.[5]

The cataclysmic Asian tsunami of 2004 and the subsequent tsunamis in 2009 and 2011 did lead to public discussions in New York City addressing disaster preparedness in the eventuality of a major oceanic disturbance. Yet, New York has done little in the way of actual storm surge management to prepare for the unthinkable scenario of sinking waterfronts and the extensive flooding of vast areas of the city's enormous coastline. Official public discussion about climate change in New York has largely centered around evacuation planning and disaster management, rather than disaster prevention, something other water-bound cities around the world have been implementing for at least thirty years. For instance, Singapore has developed a giant marina comprising a system of nine crest gates operated by giant pumps to improve drainage infrastructure and contain flood plains. Rotterdam has built a complex of seawalls, barriers, and dams designed to redirect storm surge waters to the advantage of the city. The Rotterdam model's innovation lies in its aim to convert a potentially destructive influx of storm water into usable water. Rotterdam's preparation to prevent storm surges includes a water plaza for leisure, a water-storage facility, a floodable terrace, and waterproof commercial spaces. And London has strategically constructed ten giant surge barriers along the River Thames to dissipate the force of a surge.[6]

New York, in comparison, is inexcusably behind the times when it comes to being prepared for climate change. For a city of its scale and import, it has little to offer the world, regarding how cities can improve their infrastructure. Instead, New York is an example of environmental hubris. After the superstorm, the city became emblematic of the sort of thinking that leads to the devastation wrought by climatic extremes. This thinking, that Hurricane Sandy was a once-in-a-lifetime storm, that this kind of oceanic disturbance is an anomaly that won't happen here again, is the sort of planning error that will allow future water-related disasters in New York.[7] Leaving infrastructural change to the vagaries of politicians preoccupied with how much mileage investing in expensive infrastructural building will bring them in the next elections is a dangerous path for New York.

So far, adaptation in New York to climate change has been relegated largely to cosmetic government initiatives such as bike lanes, transportation redesign, expensive architectural makeovers, and city-beautification gestures in the public sector. The city's future has not yet been viewed as an inescapable, shared fate requiring large-scale, expensive engineering redesign for storm surge protection. Instead, the conversations in the aftermath of Sandy suggest that communities and regions affected by the storm are being asked to independently deal with the environmental catastrophe as a private, personal tragedy.[8] This approach permeates many spheres of everyday life following the hurricane.

The most obvious example of the blurred lines of post-Sandy accountability was the dramatic visual impact of the electrical blackout across downtown Manhattan, from Thirty-ninth Street southward.[9] One resident of Greenwich Village described walking south from the well-lit mid-Manhattan area toward an eerily dark Greenwich Village below Fourteenth Street as an apocalyptic journey. This cloak of darkness, the result of the collapse of New York City's electrical grid, exposed the city's inadequate electrical infrastructure, which had been flooded by the storm surge. In contrast, the Netherlands has invested in a storm-proof power grid, installed underground and using water-resistant pipes. Their power lines are designed to withstand the extremes of wind and rain.[10]

Planning for Sandy ought to have included generators for every high-rise. Instead, negligence in Albany, City Hall, and at the neighborhood level left a vertical city in crisis. Large populations of people on high floors were stranded without elevators or safety lights in their emergency stairwells. The shocking case of the Sand Castle, four gargantuan tower complexes in Far Rockaway, Queens, staged the extreme horrors of poor disaster preparation and the privatization of storm surge planning. Left without electricity, water, assistance of any kind, and cut off from the outside world because of power failure, the high-rise complex became a deplorable example of the dangers of vertical living.[11] Astonishingly, very few high-rises in New York City have thought it necessary to bolster their emergency planning with the minimum of backup generators to run at least one elevator per building.

Mold in dry walls and the crevices of flooded homes also came to be treated as a personal issue, rather than a public health matter related to the storm surge. One of the consequences of flooding by seawater, seeped sewage, and the contamination of neighborhoods by untreated sewage is the problem of mold. Mold singularly contributed to the large-scale damage of household property, clothing, furniture, home infrastructure, and basement structures during Sandy. The spread of mold is swift and hard to clean up. Its toll on public health is deleterious. Moldy furniture has to be destroyed.[12]

In the wake of Sandy, the extent to which individuals and communities were able to clean their homes of mold became a question of economics. The more educated and affluent the community, the quicker the blight of mold was attended to. For underserved communities, the danger of mold considerably deteriorated their quality of life.[13] Many affected neighborhoods did not have early access to information or assistance to cope with the post-hurricane clean up. Families were temporarily displaced from residences deemed uninhabitable by city officials.[14] After a major flood, the eradication of mold to avoid environmental biohazards is the sort of public health issue, among many, that cities like New York must prepare for. Educating residents of coastal cities in public health concerns related to storm surges should be part of the larger plan for storm surge preparation.

The problem of mold is just one among many impending planning issues that Hurricane Sandy raised in the first few weeks following the storm. The panic around gasoline and transportation networks was another major point of anxiety and breakdown. Reports suggest that panic around the perceived lack of fuel led to the extreme shortage of gas in the New York City region, further exacerbating an already stressful existence with additional hours of commute and lines of waiting to get a rationed amount of gas. The shortage of gas led to a lack of food and services requiring gas to keep them running.[15] The resulting paralysis of poorly stocked grocery stores, disrupted transit, fragmentary social services, and intermittent cab services generated a slowing down of economic activity in the city.

The converging issues of electrical outage, mold, and gasoline shortage during the storm surge are instructive of the multiple secondary issues that accompany urban water disasters. They make up what in disaster management are termed "lifeline systems."[16] These communicative infrastructures integrate transit, power, and communications after a catastrophe. Hurricane Sandy brought to light the extent of unpreparedness on the part of both the state and city governments in New York, despite warnings about the scale of the storm.[17]

Water as "Blue Gold"

Much of the conversation regarding storm surges in New York assumes that water is a problem. The assumption is that the right set of solutions, such as a storm barrier, will dissipate the problem. But this is where New York can find inspiration across the Atlantic, in the maritime town of Rotterdam, Holland. In 1953, a catastrophic storm surge forced the city to start constructing barriers, seawalls, and dams. The national undertaking, titled Delta Works, was recently revamped with a rethinking of water as an opportunity, rather than a problem. Arnoud Molenaar, the head of the Rotterdam Climate Proof Program, states that the group took the approach that water from the sky and ocean could be converted into "blue gold." According to Molenaar, the conventional emphasis to block water from gushing in, or the reactive approach taken by New York City to evacuate people out of the path of the deluge, is to miss the challenge presented by the situation: how can excess water be productively harnessed for storm surge management.[18]

Molenaar's thinking on storm surge management is of immense import for New York City. The city will have to figure out ways to break the path of destructive surges and disperse the approaching fury of the ocean in the coming years. This scenario has already been presented by NYS 2100, the commission formed by Governor Andrew M. Cuomo to review and propose changes to the state's infrastructure, to prepare for further extreme weather.

Landscapes of Adaptation

Seawalls, enormous harbor-spanning surge barriers with movable gates, the reclamation of oyster beds, and building dunes, wetlands, and oyster reefs are only some of the broad array of plans being proposed by city and state officials since Hurricane Sandy. Cultivating natural defenses as well as investing in hard infrastructure is the strategy invoked for the changing landscape of New York's waterscape. A long list of infrastructural updating, including modernizing the city's sewer system, protecting the subway networks, and strengthening utility services to avoid long blackouts, is being made, despite the skepticism as to where the money will come from to pay for this restructuring. A new mayor will be elected in 2013, and climate change preparedness is emerging as a critical topic in all discussions relating to New York.[19]

But the city has very little time before the next storm surge hits, perhaps in a few years. After receiving the reports of the four commissions he charged to study the various aspects of Hurricane Sandy, Governor Cuomo emphasized in his state of the state address of 2013 that "there is a 100 year flood every two years now."[20] By explicitly locating New York City as an island city with a specific ecology of manmade environments that are vulnerable to rising sea levels, the governor has officially introduced climate change as a major part of New York's future planning initiatives. Hardening the infrastructure of New York's airports, transit networks, communication networks, and electrical, gas, and utilities networks are only some of the vast areas of restructuring the governor has brought to the official discourse of New York's future planning.

However, all these proposals remain just proposals, unless the political will in Albany and Washington shifts from rhetoric to implementing policy, before the next catastrophe hits America's financial heart.

Surviving Sandy

As a resident of the West Village who lives close to the Hudson River, I was devastated by Hurricane Sandy. During the Sandy blackout, my husband lost his footing in the pitch black emergency stairwell of our high-rise. He suffered a severe traumatic brain injury and fell into a coma of many weeks. Our life tragically came apart, as we struggled alongside thousands of other New Yorkers seeking medical help at a time when some of the city's major hospitals located downtown were shut down due to flooding. Critically ill patients were treated publicly, in open thoroughfares, corridors, and hospital lobbies in uptown Manhattan, their beds shabbily separated by makeshift cloth barriers. New York seemed more Third World than I had ever imagined. The city appeared to be in triage, unlike anything I had seen before. It stunned me to realize that our well-heeled building of twenty floors, with fancy carpets and a well-designed roof garden, did not own a generator to keep at least one elevator active after the first few hours of emergency lighting. Such an obvious oversight was incomprehensible, as I had somehow assumed that every large building had its own backup generator, as is the case in many Indian high-rises. New York's utter provincialism in emergency planning seemed incongruent with its precarious location as a coastal city in the path of major storm surges.

What surfaced during this terrible journey from personal devastation to recovery was the extraordinary social cohesiveness of friends and strangers from across New York who spontaneously stepped into my daughter's and my lives, offering solace in the desolate underworld of disaster. Neighbors and friends opened up their doors and hearts, offered to walk my dog, care for my child, shop for my groceries, as I spent long hours in the intensive care unit, week after week during Sandy's aftermath, well into Thanksgiving and Christmas. The sushi chef whose home in Queens had no water or electricity for weeks wanted to visit my husband in the hospital. The Chinese laundry staff asked about our welfare, despite the extensive hardship they were going through, without water, heat, or power for weeks in their homes, a two-hour commute away in Long Island. The celebrity butcher from Staten Island cheerfully ordered me to kick my husband back into consciousness. The extraordinary generosity of my neighbors, who cleaned my apartment, took care of my child, and left food at my door for weeks, took me by surprise, as we crawled our way through grief and shock. My bike shop mechanic made it a point to offer words of comfort in passing, amid despairing days. The Korean deli manager started giving me deals on my spring water. People rallied in unexpected ways, reminding me that my story was only one in the myriad tales of desolation and loss the city was going through.

This convivial urban sociality of formal and informal intimacies, which makes cities livable, opened up a layer of New York that is the reason many choose to brave the terrible storms that lie ahead. This intimate New York played out across the five boroughs, from the Far Rockaways and Breezy Point to Staten Island and Brooklyn. It is a New York of admirable social engagement and resilience, where people continue to look out for each other amid the city's adversities, of which there have been too many in New York's recent history.

conclusion | *toward a praxis of cosmopolitan citizenship*

New York's future lies in the reclamation of its great water resources for energy, artistic creativity, ecological sustainability, and urban futurity. Complexly divided from block to block by economics, geography, and real estate, the city is no longer the tallest in the world, or the densest. It is no longer the most iconic of contemporary cities. Yet, hemmed in by water, it offers an instructive response to David Harvey's question "By what set of institutional arrangements might all the inhabitants of planet earth hope to negotiate, preferably in a peaceful manner, their common occupancy of a finite globe?"[1] New York remains an experiment in human possibility that is continually challenging, irreverent, and people-bound in its articulations, as the protests raging on Wall Street, Foley Square, Union Square, and Washington Square in October 2011 furiously attest.

The city's most enduring lessons in urban potentiality reside in the small gestures, the passing involvements, the utopian premise of a cosmopolitan city whose very foundations are being questioned in the face of violent skirmishes between the police and citizens during the Occupy Wall Street demonstrations.

Occupy Wall Street is a generational outcry against a world gone terribly awry, a world wrought out of the very entrails of the oldest cosmopolitan American city. As the phenomenon unsettles New York City and spreads across the United States, it demands that people still matter in the face of growing disenfranchisement. In the shadow of the rising World Trade

Center Towers, hundreds of protestors downtown in Zucotti Park, formerly Liberty Park, near Wall Street, settle in for a long period of protesting. They are hopeful symbols of the struggles for cosmopolitan belonging that live despite the growing displacements of big development. These young people, defiantly weathering pepper spray and police violence, serve as reminders that the city continues to nurture imagined futures. These Wall Street protestors attest to the fact that social intimacies will continue to proliferate in ever-expanding rituals of public life, both formal and improvised, individual and communal, amid the tightening of economic mobility and the collapse of fiscal responsibility.

This book began as an investigation into the mechanisms of urbanism and the practices of citizenship-making in New York, spanning two mayors and many upheavals. It captures the disparate directions of urban growth and offers expanded notions of what participatory cosmopolitanism might look like. Despite the apocalyptic claims from many quarters about the end of cities, and the demise of optimistic scenarios for world governance, this book demonstrates a renewed interest in pursuing hopeful outcomes for global belonging. Emboldened by real politics on the streets of New York, sparked by Occupy Wall Street, a groundswell of anger and discontent resisting the hopeless scenarios of a greedy and ethically bankrupt financial culture, this book is an investigation into cosmopolitan belonging.

As a node in a larger network of global cities, New York City remains a wellspring of infinite cosmopolitan desires. It is the first building block in a larger framework for political and cultural praxis. New York's once open, now policed, waters continue to draw spirited innovators to its ever-changing shores. The city remains a working endeavor of cosmopolitan citizenship, shedding light on the future of urban imaginings.

Through the city, the more abstract arrangements of identification, such as the cosmopolitan, the world citizen, can continue to be argued for.

notes

Prologue

1. David M. Halbfinger, Charles V. Bagli, and Saah Maslin Nir, "On Ravaged Coastline, It's Rebuild Deliberately vs. Rebuild Now," *New York Times*, December 22, 2012.

2. Residents of Westbeth, an artists' colony in the West Village, reported nine feet of water in their basements.

3. Halbfinger, Bagli, and Nir, "On Ravaged Coastline, It's Rebuild Deliberately vs. Rebuild Now."

4. Matthew L. Wald and Danny Hakim, "Storm Panel Recommends Major Changes in New York," *New York Times*, January 7, 2013.

5. Eric Klinenberg, "Adaptation: How Can Cities Be 'Climate-Proofed'?" *The New Yorker*, January 7, 2013.

6. Sheri Fink, "Where Fear, Death and Myth Collided: After Storm, Web Rumors Overtook Agony at Queens High Rise," *New York Times*, December 20, 2012.

7. Michael Schwirtz, "Waste Flows after Storm Expose Costly Defects in Sewage System," *New York Times*, November 30, 2012.

8. Barry Drogin, "Surviving in the Sandy Superdome; We Were Ignored!" *The Villager*, November 22–28, 2012.

Introduction

1. Jacques Derrida, *On Cosmopolitanism and Forgiveness* (New York: Routledge, 2001), 5.

2. Russell Shorto, *The Island at the Center of the World: The Epic Story of Dutch Manhat-*

tan and the Forgotten Colony That Shaped America (New York: Doubleday, 2004). Derrida, *On Cosmopolitanism and Forgiveness*.

3. David Held, *Cosmopolitanism: Ideas and Realities* (Cambridge: Polity, 2010), 105. Also see Derrida, *On Cosmopolitanism and Forgiveness*.

4. Derrida, *On Cosmopolitanism and Forgiveness*; Emmanuel Levinas, *Beyond the Verse: Talmudic Readings and Lectures* (New York: Continuum, 2007).

5. Walter Benjamin, *Reflections: Essays, Aphorisms, Autobiographical Writings* (New York: Schocken, 1978). See Ulf Hannerz, *Exploring the City: Inquiries toward an Urban Anthropology* (New York: Columbia University Press, 1983).

6. Janet Abu-Lughod, *From Urban Village to East Village: The Battle for New York's Lower East Side* (Cambridge, MA: Blackwell, 1994).

7. See Shorto, *The Island at the Center of the World*, 6. Also see Eric W. Sanderson, *Mannahatta: A Natural History of New York City* (New York: Abrams, 2009), 17.

8. Anthony D. King, ed. *Culture, Globalization, and the World-System: Contemporary Conditions for the Representation of Identity* (Minneapolis: University of Minnesota Press, 1997). Also see the Dutch East India Company's maps of their colonial outposts, illustrated in Johannes Vingboons, *Atlas van kaarten en aanzichten van de VOC en WIC, genoemd Vingboons-atlas, in Het algemeen rijksarchief te's-Gravenhage* ([Bussum]: Fibula-Van Dishoeck, 1981).

9. Marc Augé, *In the Metro*, translated by Tom Conley (Minneapolis: University of Minnesota Press, 2002).

10. Johnny Dwyer, "Trying Times in Little Liberia," *Village Voice*, August 19, 2003.

11. Gerald Benjamin and Richard P. Nathan, *Regionalism and Realism: A Study of Governments in the New York Metropolitan Area* (Washington, DC: Brookings Institution, 2001).

12. Moustafa Bayoumi, "Letter to a G-Man," in Michael Sorkin and Sharon Zukin, eds. *After the World Trade Center: Rethinking New York City* (New York: Routledge, 2002).

13. "Sikh Outcry over Post-9/11 Bigotry," *Metro*, April 15, 2008.

14. Jasbir Puar, *Terrorist Assemblages: Homonationalism in Queer Times* (Durham, NC: Duke University Press, 2007).

15. Levinas, *Beyond the Verse*, 40.

Part I: Fluid Urbanism

1. Paolo Virno, *A Grammar of the Multitude: For an Analysis of Contemporary Forms of Life*, translated by Isabella Bertoletti and James Cascaito (Los Angeles: Semiotext(e), 2009).

Chapter 1: Water Ecology, Island City

1. "Viele's Water Map," in Paul E. Cohen and Robert T. Augustyn. *Manhattan in Maps: 1527–1995* (New York: Rizzoli, 2006), 136.

2. Eric W. Sanderson, *Mannahatta: A Natural History of New York City* (New York: Abrams, 2009).

3. Rem Koolhaas, *Delirious New York: A Retroactive Manifesto for Manhattan* (New York: Monacelli, 1997).

4. John Cronin and Robert Kennedy *The Riverkeepers: Two Activists Fight to Reclaim Our Environment as a Basic Human Right* (New York: Touchstone, 1999).

5. Cronin and Kennedy, "Battleground," in *The Riverkeepers*. Also see Sanderson, *Mannahatta*, 87.

6. "The River and the Land," in *The Hudson: A History* by Tom Lewis (New Haven, CT: Yale University Press, 2007).

7. Sanderson, *Mannahatta*. Also see Leslie Day, *Field Guide to the Natural World of New York City* (Baltimore: Johns Hopkins University Press, 2007).

8. Day, *Field Guide to the Natural World of New York City*.

9. The retaining wall that was exposed by the World Trade Center collapse at the corner of West Street and Vesey Street accentuated to passersby the constructed nature of the island's land mass.

10. Advertisements on the A train in New York City in 2009 solicited people with respiratory problems and headaches stemming from the World Trade Center disaster to seek help in New York hospitals.

11. Gerard T. Koeppel, *Water for Gotham: A History* (Princeton, NJ: Princeton University Press, 2000). Also see Mark Kurlansky, *The Big Oyster: History on the Half Shell* (New York: Random House, 2007).

12. "Building the Waterfront," in *Unearthing Gotham: The Archaeology of New York City* by Anne-Marie Cantwell and Diana diZerega Wall (New Haven, CT: Yale University Press, 2001).

13. The Mannahatta Project, under the direction of Eric W. Sanderson, of the New York City Wildlife Conservancy, has made available to the public the detailed topography of Manhattan in 1609, the time of Henry Hudson's arrival in Mannahatta.

14. Michael Sorkin and Sharon Zukin, eds. *After the World Trade Center: Rethinking New York City* (New York: Routledge, 2002).

15. Marshall Berman, "When Bad Buildings Happen to Good People," in Sorkin and Zukin, *After the World Trade Center*.

16. "The Dilemma of Waterfront Development," in *Waterfront: A Walk around Manhattan* by Phillip Lopate (New York: Anchor, 2005).

17. "The Dilemma of Waterfront Development," in *Waterfront* by Lopate.

18. Smriti Srinivas, *Landscapes of Urban Memory: The Sacred and the Civic in India's High-Tech City* (Minneapolis: University of Minnesota Press, 2001).

19. Victoria Marshall, Brian McGrath, Joel Towers, and Richard Plunz, eds., *Designing Patch Dynamics* (New York: Columbia University GSAPP, 2008).

20. New York City Department of City Planning, *Zoning Handbook* (New York, 2009).

21. Anthony Flint, *Wrestling with Moses: How Jane Jacobs Took On New York's Master Builder and Transformed the American City* (New York: Random House, 2009).

22. Jotham Sederstrom, "Ruling Could Put Ratner's Atlantic Yards Project Back on Track," *Daily News* (Brooklyn), May 16, 2009.

Chapter 2: Transoceanic New York, City of Rivers

1. "Colman Slaine. Colmans Point. Treacherous Sausages," in *Juet's Journal of Hudson's 1609 Voyage, from the 1625 Edition of Purchas His Pilgrimes*, Robert Juet, transcribed by Brea Barthel, accessed October 10, 2012, http://www.halfmoon.mus.ny.us/Juets-journal.pdf, p. 13. Eric W. Sanderson locates Henry Hudson's *Half Moon* off of Greenwich Village on September 12, 1609. See Eric W. Sanderson, *Mannahatta: A Natural History of New York City* (New York: Abrams, 2009), 293.

2. Russell Shorto's book *The Island at the Center of the World: The Epic Story of Dutch Manhattan and the Forgotten Colony That Shaped America* (New York: Doubleday, 2004) paved the way for thinking about the considerable Dutch legacies in New York culture. Also see Jaap Jacobs, *The Colony of New Netherland: A Dutch Settlement in Seventeenth-Century America* (Ithaca, NY: Cornell University Press, 2009); Martine Gosselink, *New York New Amsterdam: The Dutch Origins of Manhattan* (New Amsterdam: Nieuw Amsterdam, 2009); Joseph O'Neill, *Netherland* (New York: Pantheon, 2008).

3. Shorto, *The Island at the Center of the World*, 16, 17.

4. Petrus Plancius, *Map of the World*, 1596, paper, Collection National Archief (National Archives of the Netherlands), in *New Amsterdam: The Island at the Center of the World*, South Street Seaport Museum, September 2009.

5. "Colman Slaine. Colmans Point. Treacherous Savages," in *Juet's Journal of Hudson's 1609 Voyage*, 13.

6. Verenigde Oostindische Compagnie (VOC), Dutch acronym for the Dutch East India Company.

7. Museum of the City of New York, *Amsterdam/New Amsterdam: The Worlds of Henry Hudson*, curated by Susan Henry, 2009.

8. Museum of the City of New York, *Amsterdam/New Amsterdam*.

9. Arnoldus Montanus, *The New and Unknown World: Descriptions of America and the South Lands*. Koninklijke Bibliotek, the Hague, Netherlands, 185 B 14.

10. Jan Huygen Van Linschoten, *The Voyage to the East Indies*. Volumes 1 and 2 (London: Elibron Classics, 2005), 1:175, 1:183, 1:277, 1:279, 2:79, 2:115, 2:119.

11. Van Linschoten, *The Voyage to the East Indies*; Adriaen Van Der Donck, *A Description of New Netherland* (Lincoln: University of Nebraska, 2008); Johannes De Laet, *Extracts from the New World or a Description of the West Indies* (Ithaca, NY: Cornell Uni-

versity Library Digital Collections, 1993); David Pietersz De Vries, "Voyages from Holland to America, A.D. 1633 to 1644," in New-York Historical Society, *Collections of the New-York Historical Society. Volume 111–Part 1* (New York: D. Appleton and Company, 1857), 96, 97, 106.

12. Isaack De Rasieres, *De Rasieres' Letter*, in *New Netherland in 1627* (Ithaca, NY: The Cornell University Library Digital Collections, 1993), 345.

13. De Rasieres, *De Rasieres' Letter*, 346 (author's emphasis).

14. Johannes Megapolensis, *A short sketch of the Mohawk Indians in New Netherland: their land, stature, dress, manners, and magistrates, written in the year 1644, by Johannes Megapolensis, and notes, by John Romeyn Brodhead* (Ithaca, NY: Cornell University Library Digital Collections, 1993), 155.

15. De Vries, "Voyages from Holland to America," in *Collections of the New-York Historical Society, Volume 111*, 3:96, 97, 106. Also see Edmund Bailey O'Callaghan, *History of New Netherland or New York under the Dutch. Vol. 1* (New York: Elibron Classics, 2005), 61, 62, 64, 65.

16. J. Nieuhoff, *Voyages and Travels into Brazil and the East-Indies, 1640–1649* (New Delhi: Asian Educational Services, 2003).

17. De Vries, quoted in J. Franklin Jameson, ed. *Narratives of New Netherland, 1609–1664* (New York: Elibron Classics, 2005), 189, 191, 193.

18. De Vries, quoted in J. Franklin Jameson, ed. *Narratives of New Netherland*, 43.

19. De Vries, quoted in J. Franklin Jameson, ed. *Narratives of New Netherland*, 41.

20. De Vries, quoted in J. Franklin Jameson, ed. *Narratives of New Netherland*, 188, 189.

21. Van Linschoten, *Travels to the Portuguese East Indies*; Hans Staden, "The True History and Description of a Country Populated by a Wild, Naked, and Savage Man-munching People, situated in the New World, America . . ." in *Hans Staden's True History: An Account of Cannibal Captivity in Brazil*, edited and translated by Neil L. Whitehead and Michael Harbsmeier (Durham, NC: Duke University Press, 2008). Staden was a German mariner who wrote his account of Portuguese Brazil in 1557.

22. Johannes Vingboons, *Atlas van kaarten en aanzichten van de VOC en WIC, genoemd Vingboons-atlas, in Het algemeen rijksarchief te's-Gravenhage* ([Bussum]: Fibula-Van Dishoeck, 1981). Through his elaborate cartographic renderings of the numerous ports of the Dutch East India Company, Vingboons threads an interconnectivity that is extensive in scope and scale, strung across the globe.

23. Johannes Vingboons, *New Amsterdam or New York on the Island Man*, watercolor, c. 1665, Collection National Archief 4, VELH 619–014.

24. Jan Morris, *The Great Port: A Passage through New York* (New York: Oxford University Press, 1985). Also see William Kornblum, *At Sea in the City: New York from the Water's Edge* (Chapel Hill, NC: Algonquin, 2002); Brian Cudahy, *Around Manhattan Island and Other Tales of Maritime New York* (New York: Fordham University Press, 1997); *Wiley and Putnam's Emigrant's Guide: Comprising Advice and Instruction in Every Stage*

of the Voyage to America (London: Wiley and Putnam, 1845); Rem Koolhaas, *Delirious New York: A Retroactive Manifesto for Manhattan* (New York: Monacelli, 1997).

25. "Lenape Country and New Amsterdam to 1664," in Edwin G. Burrows and Mike Wallace, *Gotham: A History of New York City to 1898* (New York: Oxford University Press, 2009), 24.

26. "Lenape Country and New Amsterdam to 1664," in Burrows and Wallace, *Gotham.*

27. De Vries, "My Third Voyage to America and the New Netherland," in New-York Historical Society, *Collections of the New York Historical Society, Volume III*. The journals of De Vries are explicit about the brutality of the Dutch in New Amsterdam toward Native Americans. De Vries appears to have nurtured friendly relations with the Mohawk peoples and been horrified by Dutch massacres of Indian communities living around Mannahatta and its vicinity.

28. Fernand Braudel, *The Mediterranean and the Mediterranean World in the Age of Philip II* (Berkeley: University of California Press, 1995).

29. Joseph Berger, *The World in a City: Traveling the Globe through the Neighborhoods of the New New York* (New York: Ballantine, 2007).

30. "From Mahicantuck to the Millennium," in *The State of the Hudson 2009*, Hudson River Estuary Program, Department of Environmental Conservation, accessed October 10, 2012, http://www.dec.ny.gov/lands/51492.html.

31. "Hudson River Basics," in *The State of the Hudson 2009*.

32. Gerald T. Koeppel, *Water for Gotham: A History* (Princeton, NJ: Princeton University Press, 2000).

33. Isaack De Rasieres, *New Netherland in 1627* (Ithaca: Cornell University Library Digital Collections, 1993).

34. De Vries, quoted in Jameson, *Narratives of New Netherland*, 188.

35. *Amsterdam/New Amsterdam: The Worlds of Henry Hudson*, the Museum of the City of New York, 2009.

36. Thanks to Victoria Marshall for discussions on navigational maps.

37. Van der Donck, *A Description of New Netherland*, 10.

38. Cortelyou's map is found in Paul E. Cohen and Robert T. Augustyn, *Manhattan in Maps: 1527–1995* (New York: Rizzoli, 2006), 42–43.

39. Tom Lake, *Hudson River Almanac*, September 15–September 22, 2009, Hudson River Estuary Program, New York State Department of Environmental Conservation, 2007–2009.

40. Tom Lake, *Hudson River Almanac*.

41. Joan Vingboons, *Noort Rivier in Niew Neerlandt*, 1639, pen and ink, watercolor described in Richard W. Stephenson, "The Henry Harisse Collection, 40," *Terrae Incognitae: The Journal for the History of Discoveries*, vol. xvi (1984).

42. Van der Donck, *A Description of New Netherland*, 15.

43. Van der Donck, *A Description of New Netherland*, 12.

44. John Cronin and Robert Kennedy, *The Riverkeepers: Two Activists Fight to Reclaim Our Environment as a Basic Human Right* (New York: Touchstone, 1999), 21.

45. Hugh Macatamney, *Cradle Days of New York* (New York: Drew and Lewis, 1909).

46. Mark Kurlansky, *The Big Oyster: History on the Half Shell* (New York: Random House, 2007), 249–79.

47. "Enduring Shellfishness," in Kurlansky, *The Big Oyster*.

48. "Enduring Shellfishness," in Kurlansky, *The Big Oyster*.

49. Charles Duhigg, "Sewers at Capacity, Waste Poisons Waterways," *New York Times*, November 23, 2009.

50. Thanks to Daniel Hetteix for pointing this out to me.

Chapter 3: The Maritime Sky of Manhattan

1. Joyce Hansen and Gary McGowan, *Breaking Ground, Breaking Silence: The Story of New York's African Burial Ground* (New York: Holt, 1998), 17, 21, 22. See also Christopher Moore, "A World of Possibilities: Slavery and Freedom in Dutch New Amsterdam," in Berlin and Harris, *Slavery in New York*.

2. Edwin G. Burrows and Mike Wallace, *Gotham: A History of New York City to 1898* (New York: Oxford University Press, 2009).

3. Jaap Jacobs, *The Colony of New Netherland: A Dutch Settlement in Seventeenth-Century America* (Ithaca, NY: Cornell University Press, 2009), 21.

4. *Juet's Journal of Hudson's 1609 Voyage, from the 1625 Edition of Purchas His Pilgrimes*, Robert Juet, transcribed by Brea Barthel, accessed October 10, 2012, http://www.halfmoon.mus.ny.us/Juets-journal.pdf, p. 592.

5. *Juet's Journal of Hudson's 1609 Voyage, from the 1625 Edition of Purchas His Pilgrimes*.

6. Robert Juet writes of the death of John Colman, one of Hudson's shipmates, who is killed in an encounter between the Native Americans and Hudson's crew. This incident unsettles Hudson's enterprise.

7. Museum of the City of New York, *Amsterdam/New Amsterdam: The Worlds of Henry Hudson*, exhibit, April 4–September 27, 2009.

8. Russell Shorto, *The Island at the Center of the World: The Epic Story of Dutch Manhattan and the Forgotten Colony That Shaped America* (New York: Doubleday, 2004).

9. Shorto, *The Island at the Center of the World*, 56–57.

10. Shorto, *The Island at the Center of the World*, 57.

11. Oliver A. Rink, *Holland on the Hudson: An Economic and Social History of Dutch New York* (Ithaca, NY: Cornell University Press, 1986).

12. Colonel Egbert L. Viele's 1865 map of Manhattan outlines the rivers and underground rivulets that flow through the built, and still exposed, areas of Manhattan. See also Eric W. Sanderson's *Mannahatta Project*. The Wildlife Conservation Society is simulating the topography of the island of Manhattan as possibly encountered by Henry Hudson in September 1609.

13. Jacobs points out that the passenger lists on board the ships tended to ignore women and children along with enslaved black individuals. Furthermore, the few entries available in the archives of early Dutch arrivals to New Amsterdam only cursorily indicate the ethnicity of the domestic labor accompanying immigrating European populations.

14. David Pietersz De Vries, *Korte historiael ende journal notes*, 1655, National Maritime Museum, Amsterdam. De Vries (1592/1593–1655) made three voyages to New Netherlands in the 1630s and 1640s. He stayed for ten years and published his travels to the Mediterranean, Newfoundland, France, and later the East Indies. It was not uncommon for employees of the Dutch East India Company to move within their colonies across hemispheres for reasons of work and sheer curiosity, as the writings of De Vries suggest.

15. Ira Berlin and Leslie Harris, eds., *Slavery in New York* (New York: New Press, 2005), 34.

16. Berlin and Harris, *Slavery in New York*, 33.

17. Berlin and Harris, *Slavery in New York*, 6, 43.

18. Ira Berlin and Leslie Harris, "Uncovering, Discovering, and Recovering: Diggin in New York's Slave Past Beyond the African Burial Ground," *New York Journal of American History* 66, no. 2 (fall–winter, 2005): 26–27. Also see Berlin and Harris, *Slavery in New York*, 41–46.

19. New-York Historical Society, *Slavery in New York*, exhibit, October 7, 2005.

20. Saidiya V. Hartman, *Scenes of Subjection: Terror, Slavery, and Self-Making in Nineteenth-Century America* (New York: Oxford University Press, 1997).

21. Against vociferous neighborhood opposition, Trump Development proceeded with excavating the site, which revealed 150-year-old bones. The area was part of an abolitionist church in the nineteenth century. "Fatal Collapse at Trump Soho Condo-Hotel," *Greenwich Village Society for Historic Preservation*, January 15, 2008.

22. Le Corbusier, quoted in Philip Kasinitz, *Metropolis: Center and Symbol of Our Times* (New York: New York University Press, 1995), 103.

23. Aihwa Ong, *Buddha Is Hiding: Refugees, Citizenship, the New America* (Berkeley: University of California Press, 2003).

24. Jane Jacobs, *The Nature of Economies* (New York: Vintage, 2000),

25. Thomas Bender, *The Unfinished City: New York and the Metropolitan Idea* (New York: New York University Press, 2007).

26. Michel De Certeau, *The Practice of Everyday Life* (Berkeley: University of California Press, 1984).

27. Tony Hiss, *The Experience of Place: A New Way of Looking at and Dealing with Our Radically Changing Cities and Countryside* (New York: Vintage, 1991).

28. Ulf Hannerz, *Transnational Connections: Culture, People, Places* (New York: Routledge, 1996).

29. Erving Goffman, *The Presentation of Self in Everyday Life* (New York: Anchor, 1959).

30. Jane Jacobs, *The Death and Life of Great American Cities* (New York: Random House, 1961).

31. Martin Heidegger, *Poetry, Language, Thought*, translated by Albert Hofstadter (New York: Harper and Row, 1971), 161.

32. Kenneth Frampton and John Cava, eds., *Studies in Tectonic Culture: The Poetics of Construction in Nineteenth and Twentieth Century Architecture* (Cambridge, MA: MIT Press, 1995).

33. Bernard Tschumi, *Architecture and Disjunction* (Cambridge, MA: MIT Press, 1996). Also see Bernard Tschumi, *Event-Cities* (Cambridge, MA: MIT Press, 1994).

34. Jerilou Hammett and Kingsley Hammett, eds., *The Suburbanization of New York: Is the World's Greatest City Becoming Just Another Town?* (New York: Princeton Architectural Press, 2007).

Chapter 4: Thinking Metropolitanism

1. Thomas Bender, *The Unfinished City: New York and the Metropolitan Idea* (New York: New York University Press, 2007), 224.

2. Saskia Sassen, *The Global City: New York, London, Tokyo* (Princeton, NJ: Princeton University Press, 2001), chapters 7, 8, 9.

3. Sassen, *The Global City*, 260, 261, 300, 301.

4. Le Corbusier, "New York Is Not a Completed City," in Philip Kasinitz, *Metropolis: Center and Symbol of Our Times* (New York: New York University Press, 1995).

5. Jean Gottman, *Megalopolis: The Urbanized Northeastern Seaboard of the United States* (Cambridge, MA: MIT Press, 1964); Edward W. Soja, *Postmetropolis: Critical Studies of Cities and Regions* (Malden, MA: Blackwell, 2000), 259. Also see Jonathan Barnett, *The Fractured Metropolis: Improving the New City, Restoring the Old City, Reshaping the Region* (New York: HarperCollins, 1995); Joel Garreau, *Edge City: Life on the New Frontier* (New York: Doubleday, 1991).

6. Skyscraper Museum, New York City, *Vertical Cities: Hong Kong, New York*, exhibit, 2008.

7. The Commissioner's Grid of 1811.

8. Anthony Flint, *Wrestling with Moses: How Jane Jacobs Took On New York's Master Builder and Transformed the American City* (New York: Random House, 2009). Also see Robert A. Caro, *The Power Broker: Robert Moses and the Fall of New York* (New York: Random House, 1975).

9. Le Corbusier, "New York Is Not a Completed City."

10. Jane Jacobs, *The Death and Life of Great American Cities* (New York: Random House, 1961).

11. The Brooklyn waterfront developments proposed by the giant conglomerate Forest City Ratner and Hudson River Park Trust, from the West Side of Manhattan, are two examples of how the conversation about energy potential and the city's ability to sustain its growth is being contested.

In the case of the Brooklyn megablock transformation of a historic waterfront mixed-income neighborhood to a giant condominium village for high-income residents, the people of Brooklyn put up much struggle and resistance to the impending development, whose imperative is to compete with the skyline of Manhattan. The Hudson River Park Trust, on the other hand, was an undertaking developing five acres of pier and waterfront space for public use on the west side of Manhattan. Both instances demonstrate how the people of New York City continue to grapple with emerging new challenges in urban living.

12. Bender, *The Unfinished City*.

13. The African burial ground is a good example of this collision between local identities and public memory. The current corporate, civic, and commercial identity of Duane Street was in conflict with the demands of New York's African American community to create a national memorial to the nearly twenty thousand African Americans buried in the area.

14. Herbert J. Gans, *The Levittowners: Ways of Life and Politics in a New Suburban Community* (New York: Random House, 1967).

15. Edward W. Soja, *Postmetropolis: Critical Studies of Cities and Regions* (Malden, MA: Blackwell, 2000), 218.

16. Thanks to Sikivu Hutchinson for her research on the Los Angeles transportation system.

17. Edward W. Soja, *Thirdspace: Journeys to Los Angeles and Other Real-and-Imagined Places* (Cambridge, MA: Blackwell, 1996), 239.

18. Sassen, *The Global City*, 260, 261.

19. *World Urbanization Prospects: The 2009 Revision. File 11a: The 30 Largest Urban Agglomerations Ranked by Population Size at each point in time, 1950–2025*, POP/DB/WUP/Rev.2009/2/F11a, United Nations, Department of Economic and Social Affairs, Population Division.

20. Soja, *Postmetropolis*, 250.

21. Gerald Benjamin and Richard P. Nathan, *Regionalism and Realism: A Study of Governments in the New York Metropolitan Area* (Washington, DC: Brookings Institution, 2001), 91.

22. Brian McGrath's video installation *Timeline* (2002) captures the growth of New York City from nineteenth-century urban core to the postmetropolis of the present moment. Many thanks to Brian McGrath for the extended conversations on urbanism and architecture that fueled the original idea for this book. Also see Brian McGrath and Jean Gardner, *Cinemetrics: Architectural Drawing Today* (Hoboken, NJ: Wiley-Academy, 2007).

23. Gottman, *Megalopolis*.

24. Allen J. Scott and Edward W. Soja, eds., *The City: Los Angeles and Urban Theory at the End of the Twentieth Century* (Berkeley: University of California Press, 1996).

25. Soja, *Postmetropolis*, 218, 219.

26. Kevin Lynch, *The Image of the City* (Cambridge, MA: MIT Press, 1960).

27. "How the Metropolis Split Apart," in Jonathan Barnett, *The Fractured Metropolis: Improving the New City, Restoring the Old City, Reshaping the Region* (New York: HarperCollins, 1995).

28. Thanks to Sikivu Hutchinson for her work on the Los Angeles rail system.

29. "The Global City," in Joseph Berger, *The World in a City: Traveling the Globe through the Neighborhoods of the New New York* (New York: Ballantine, 2007).

30. *China Prophecy: Shanghai Skyscraper Museum*, New York City, 2010. Also *Vertical Cities: Hong Kong/New York*, Skyscraper Museum, New York City, exhibit, 2009.

31. Corey Robin, *Fear: The History of a Political Idea* (New York: Oxford University Press, 2006), 31.

32. The City Project, *FY2000 Alterbudget Agenda. Quality of Life for Whom?* (May 1999) City Project, New York City, www.cityproject.org. The City Project collaborated with twenty-five other organizations to gather the documentation made available in their publication regarding the policies of Mayor Giuliani's administration.

33. Sassen, *The Global City*; Soja, *Postmetropolis*; Scott and Soja, *The City*.

34. Rudolph W. Giuliani, "Personal Responsibility and Work Opportunity Reconciliation Act of 1996," September 11, 1996, accessed October 10, 2012, www.nyc.gov/html/rwg/html/96/welfare.html.

35. The Hudson River Park project was the result of two decades of activism and neighborhood involvement. Its full realization during Bloomberg's tenure fed the illusion of Bloomberg's being the enabler of the west side of Manhattan's greening initiatives; however, the impetus was initiated by his predecessor, Mayor Giuliani, and Governor Pataki.

36. "Mr. Bloomberg's Gloomy Budget," *New York Times*, May 2, 2009.

37. Soja, *Postmetropolis*, 250.

38. Soja, *Postmetropolis*, 250.

Part II: Cosmopolitan Frugality

1. Garrett Wallace Brown and David Held, *The Cosmopolitanism Reader* (Cambridge, MA: Polity, 2010), 17.

2. Jacques Derrida, *On Cosmopolitanism and Forgiveness* (New York: Routledge, 2001).

Chapter 5: Nomadic Urbanism and Frugality

1. Joyce Hansen and Gary McGowan, *Breaking Ground, Breaking Silence: The Story of New York's African Burial Ground* (New York: Holt, 1998), 26.

2. Jeff Zeleny, "Obama Rallies Huge Crowd in New York," *The Caucus* (blog), *New York Times*, September 27, 2007, http://thecaucus.blogs.nytimes.com/2007/09/27/obama-rallies-huge-crowd-in-new-york. Also see Jason Horowitz, "Obama Campaigns Like a New Yorker," *New York Observer*, September 25, 2007.

3. Paul Gilroy, *There Ain't No Black in the Union Jack: The Cultural Politics of Race and Nation* (Chicago: University of Chicago Press, 1991). Also see James Clifford, "Notes on Theory and Travel," *Inscriptions* 5 (1989): 177–85.

4. Ulf Hannerz, *Exploring the City: Inquiries toward an Urban Anthropology* (New York: Columbia University Press, 1983).

5. Gilles Deleuze and Félix Guattari, *A Thousand Plateaus: Capitalism and Schizophrenia*, translated by Brian Massumi (Minneapolis: University of Minnesota, 1987).

6. Teshome Gabriel, "Theses on Memory and Identity: In Search of the Origin of the River Nile," *Emergences* 1 (fall): 131–37.

7. Gabriel, "Theses on Memory and Identity."

8. Paul Virilio, *Speed and Politics*, translated by Marc Polizzotti (Los Angeles: Semiotext(e), 2007).

9. Max Weber, *The City* (New York: Free Press, 1968), 173.

10. Weber, *The City*, 173.

11. Weber, *The City*, 163.

12. Don Martindale, "Prefatory Remarks: The Theory of the City," in Weber, *The City*, 53.

13. Weber, *The City*, 55.

14. Henri Pirenne, *Medieval Cities: Their Origins and the Revival of Trade* (Princeton, NJ: Princeton University Press, 1952), 39; Weber, *The City*, 55.

15. Charles Baudelaire, *The Painter of Modern Life and Other Essays* (London: Phaidon, 1995), 14.

16. Baudelaire, *The Painter of Modern Life and Other Essays*.

17. Baudelaire, *The Painter of Modern Life and Other Essays*, 19.

18. Ferdinand Tönnies, *Community and Society* (New Brunswick, NJ: Transaction, 1988), 239.

Chapter 6: Nyerere, the Dalai Lama, Gandhi

1. William Edgett Smith, *Nyerere of Tanzania* (Harare: Zimbabwe Publishing House, 1981), 178.

2. Kevin Lynch, *The Image of the City* (Cambridge, MA: MIT Press, 1960), 1.

3. Miriam Greenberg, *Branding New York: How a City in Crisis Was Sold to the World* (New York: Routledge, 2008).

4. Tony Hiss, *The Experience of Place: A New Way of Looking at and Dealing with Our Radically Changing Cities and Countryside* (New York: Vintage, 1991).

5. François Rabelais, *Gargantua and Pantagruel* (New York: Penguin Books, 2006), 125, 133, 150.

6. Geoffrey A. Barborka, *H. P. Blavatsky, Tibet and Tulku* (Adyar, India: Theosophical House, 1974), 113.

7. "The Material Question," in G. I. Gurdjieff, *Meetings with Remarkable Men* (New York: Dutton, 1963).

8. Frederick M. Alexander, *Alexander Technique: Original Writings of F.M. Alexander*, edited by Danny McGowan (Burdett, NY: Larson, 1997); Moshé Feldenkrais, *Awareness through Movement: Health Exercises for Personal Growth* (New York: Harper and Row, 1977).

9. See *The Theosophical Movement, 1875–1950* (Los Angeles: Cunningham, 1951), 311.

10. Michele H. Bogart, *Public Sculpture and the Civic Ideal in New York City, 1890–1930* (Chicago: University of Chicago Press, 1989), 3–4.

11. Alex Vitale, *City of Disorder: How the Quality of Life Campaign Transformed New York Politics* (New York: New York University Press, 2009).

12. John Eligon, "Sharpton and Seven Others Guilty in Sean Bell Protest," *New York Times*, October 8, 2008.

13. Anne Barnard, "Reliving the Sean Bell Case by Renaming a Street," *New York Times*, April 9, 2009.

14. Jane Jacobs, *The Death and Life of Great American Cities* (New York: Random House, 1961), 283.

Part III: Ecological Expressivit

1. Owen, David. *Green Metropolis: Why Living Smaller, Living Closer, and Driving Less Are the Keys to Sustainability* (New York: Riverhead, 2009), 46, 47.

2. Held, David. *Cosmopolitanism: Ideas and Realities* (Cambridge: Polity, 2010), 69.

3. Emmanuel Kant, *Toward Perpetual Peace and Other Writings on Politics, Peace, and History* (New Haven, CT: Yale University Press, 2006), 82.

4. Held, *Cosmopolitanism*, 74.

5. Held, *Cosmopolitanism*, 69.

6. Martha C. Nussbaum, "Kant and Cosmopolitanism," in *The Cosmopolitanism Reader*, ed. Garrett Wallace Brown and David Held (Cambridge, MA: Polity, 2010), 42.

7. Held, *Cosmopolitanism*, 68.

Chapter 7: Greening Hardscape

1. Bronislaw Geremek, *Poverty: A History*, translated by Agnreszka Kolakowska (Cambridge, MA: Blackwell, 1994), 248.

2. New York City Garden Coalition, press release no. 4, July 30, 1998.

3. Lincoln Anderson, "Auction Disrupted, but Charas Is Sold," *The Villager*, July 22, 1998.

4. Perez was murdered by drug dealers in Queens, in April of 1999, in retaliation for his efforts to clean up the neighborhood.

5. Critics of the High Line project bemoan the loss of multiple informal gardens that had sprung up across the abandoned rail line over its thirty years of neglect, creating a unique instance of New York's biodiversity through an ecosystem that

had been left untampered with for three decades. An anonymous sculpture garden and other anonymous artistic interventions along the rail tracks were demolished to make way for the high-concept public park.

6. New York City Garden Coalition, press release no. 4, July 30, 1998.

7. New York City Garden Coalition, press release no. 4, July 30, 1998.

8. "For Centennial, New York City Taxis to Become Canvas for 30,000 Painters," *Associated Press*, February 16, 2007.

9. Amy Zimmer, "Verdant Underground: MTA Has Green Visions," *Metro*, April 15, 2008.

10. "Group Calls for Subway Bike Racks," Mathew Sweeney, April 22, 2008, www.amNY.com/local.

11. "Big Apple Greener than Ever," Marlene Naanes, April 22, 2008, www.amNY.com/local.

12. Municipal Art Society, New York, *Jane Jacobs and the Future of New York*, exhibit, September 25, 2007.

13. Phillip Lopate, *Waterfront: A Walk around Manhattan* (New York: Anchor, 2005).

14. Joyce Purnick and Rinker Buck, "Hugh Carey's Mass-Transit Fiasco," *New York Magazine*, October 8, 1979, 12 (39): 12–15.

15. Patrick McGeehan, "Moving Beyond Its Feisty Roots, Hudson River Park Group Focuses on Fund-Raising," *New York Times*, June 20, 2011.

16. Tom Topousis and Rich Calder, "Hudson River Park under Pier Pressure," *New York Post*, December 15, 2009.

17. Stuart Waldman and Zack Winestine, *Maritime Mile: The Story of the Greenwich Village Waterfront* (New York: Mikaya, 2002).

18. Martin Manalansan, "Race, Violence, and Neoliberal Spatial Politics in the Global City," *Social Text* 23, no. 3–4 (fall–winter 2005): 84–85.

19. The northern part of the High Line is still owned by the CSX Railway Corporation or the Metropolitan Transit Authority.

Chapter 8: Marathon City, Biking Boroughs

1. Liz Robbins, *A Race Like No Other: 26.2 Miles through the Streets of New York* (New York: HarperCollins, 2008). Also see Haruki Murakami, *What I Talk about When I Talk about Running: A Memoir* (New York: Knopf, 2008), and Jean Echenoz, *Running: A Novel* (New York: New Press, 2009).

2. An offshoot of the emergence of cycling as a serious urban engagement is the subcultural Bike Kill Festival, held in New York City annually. A combination of sporting event, contest, carnival, performance, cult gathering, and style parade, the festival is a violent, cathartic embrace of biking lifestyles that draws the most innovative and creative two-wheel innovations.

3. Sewell Chan, "Police Investigate Officer in Critical Mass Video," *New York*

Times, July 28, 2008. Also see Jen Benepe, "Can Critical Mass Negotiate a Truce?" *Gotham Gazette*, November 22, 2004.

4. Amy Goodman, "Guantanamo on the Hudson: Detained RNC Protesters Describe Prison Conditions," *Democracy Now*, September 2, 2004.

5. "NYC DOT—DOT Completes Unprecedented Three-Year, 200 mile Installation of Bike Lanes," Press Release no. 09–030, Wednesday, July 8, 2009. See NYC .gov website.

6. "NYC DOT—DOT Completes Unprecedented Three-Year, 200 mile Installation of Bike Lanes," Press Release no. 09–030, Wednesday, July 8, 2009. See NYC .gov website.

7. "NYC DOT—DOT Completes Unprecedented Three-Year, 200 mile Installation of Bike Lanes," Press Release no. 09–030, Wednesday, July 8, 2009. See NYC .gov website.

Chapter 9: Brooklyn Carnival and the Sale of Dreamland

1. Benedict Anderson, *Imagined Communities: Reflections on the Origin and Spread of Nationalism* (New York: Verso, 1991). In his study of the rise of nation-states, Anderson identifies the nationalist sentiments for homeland experienced in dispersal as long-distance nationalism.

2. The work of British sociologists like John Clarke, Errol Lawrence, Stuart Hall, Kobena Mercer, and Brian Roberts, as well as cultural producers like Mustapha Matura, Isaac Julien, Shani Mootoo, and Maureen Blackwood, explore the meanings of carnival in Britain and the Caribbean.

3. "Haitians, Demanding Justice, Rally in New York," *U.S. News*, August 23, 1997, 2.

4. Greg Sargent, "The Incredibly Bold, Audaciously Cheesy, Jaw-Dropping Vegasified, Billion-Dollar Glam-Rock Makeover of Coney Island," *New York Times*, September 18, 2005.

5. Mayor's Office of Operations, Office of Environmental Coordination, *Coney Island Rezoning Plan: Technical Memorandum 002*, issued July 22, 2009.

6. Amanda Fung, "Coney Island Keeper," *Crain's New York Business*, June 28, 2009, 8.

7. Omar Robau, "Thor Equities Announces Their Land Fully Active This Summer," *Kinetic Carnival*, April 20, 2009. www.Kineticcarnival.blogspot.com/2009/04/thor-equities-announces-their-land.html.

Chapter 10: Spirits of the Necropolis, Planes on the Hudson

1. Martin Manalansan, "Race, Violence, and Neoliberal Spatial Politics in the Global City," *Social Text* 23, nos. 3–4 (fall–winter, 2005): 84–85.

2. Rick Beard and Leslie Cohen Berlowitz, *Greenwich Village: Culture and Counterculture* (New Brunswick, NJ: Rutgers University Press, 1993), 40. Also see Stuart Wald-

man and Zack Winestine, *Maritime Mile: The Story of the Greenwich Village Waterfront* (New York: Mikaya, 2002).

3. Thanks to Daniel Hetteix for pointing this out to me.

Chapter 11: Governors Island

1. Emmanuel Levinas, *Beyond the Verse: Talmudic Readings and Lectures* (New York: Continuum, 2007, 34.

2. Susan L. Glen, *Images of America: Governors Island, NY* (Charleston, SC: Arcadia, 2009), 9.

3. Diana diZerega Wall and Anne-Marie Cantwell, *Touring Gotham's Archaeological Past: 8 Self-Guided Walking Tours through New York City* (New Haven, CT: Yale University Press, 2004), 8.

4. Glen, *Images of America*.

5. Glen, *Images of America*, 12.

6. Wall and Cantwell, *Touring Gotham's Archaeological Past*, 9.

7. Wall and Cantwell, *Touring Gotham's Archaeological Past*, 38.

8. Glen, *Images of America*, 80.

9. The extensive documentation by the Dutch cartographer Johannes Vingboons archives many of the VOC's fort cities across its Asian and African empire. The cities of Cochin, Quilon, and Galle, in South India and Sri Lanka, are cases of Dutch fort cities that later became important ports. Cape Town, South Africa, and Recife, the capital of Pernambuco, Brazil, were some of the more prominent fort cities built by the Dutch, along with New Amsterdam in Curacao and New Amsterdam in Mannahatta.

10. Ann-Marie Cantwell and Diana diZerega Wall, *Unearthing Gotham: The Archaeology of New York City* (New Haven, CT: Yale University Press).

11. Glen, *Images of America*, 48.

12. The VOC had colonial outposts all across Asia and Africa. Fort Cochin in Kerala, India, my parents' home and my former home, is one of them.

13. Museum of the City of New York, *Amsterdam/New Amsterdam. The Voyages of Henry Hudson*, exhibit, 2009.

14. While the exhibits curated around New York State on the subject of Dutch New York are too many to cite here, the following exhibits are of particular note: Crailo State Historic Site, *New Netherland: A Sweet and Alien Land*; Hudson River Museum, *Dutch New York: The Roots of Hudson Valley Culture*; Albany Institute of History and Art, *Hudson River Panorama: 400 Years*; Hudson River Valley Heritage, *Hudson-Fulton Celebration of 1909*; Thomas Cole Historic Site, *River Views of the Hudson River School*; Museum of the City of New York, *Amsterdam/New Amsterdam: The World of Henry Hudson*; New York State Museum, *1609*; Wave Hill, Bronx, *The Muhheakantuck in Focus*.

15. Brochure of NY400 Holland on the Hudson presents *New Island Festival September 10–20 Governors Island*.

16. According to Wall and Cantwell, the evidence of a Native American hunting presence on Governors Island has been unearthed, extending to 4,000 years ago. Wall and Cantwell, *Touring Gotham's Archaeological Past.*

17. David Held, *Cosmopolitanism: Ideas and Realities* (Cambridge: Polity, 2010), 74.

18. Museum of the City of New York, *Amsterdam/New Amsterdam: The World of Henry Hudson*, exhibit, 2009.

Chapter 12: After Hurricane Sandy

1. N. R. Kleinfield, "Battered Seaside Haven Recalls Its Trial by Fire," *New York Times*, December 25, 2012.

2. Lincoln Anderson, "Quinn Floats Raft of Ideas for Fighting Future Floods," *The Villager*, November 15–21, 2012.

3. "The Power Mess on Long Island," *New York Times*, November 19, 2012.

4. Eric Klinenberg, "Adaptation: How Can Cities Be 'Climate-Proofed'?" *The New Yorker*, January 7, 2013, 33. Also see "Transcript of Governor Andrew M. Cuomo's 2013 State of the State Address," Andrew M. Cuomo, Albany, N.Y., January 9, 2013, www.governor.ny.gov/press/01092013sostranscript.

5. Matthew L. Wald and Danny Hakim, "Storm Panel Recommends Major Changes in New York," *New York Times*, January 7, 2013.

6. Anderson, "Quinn Floats Raft of Ideas for Fighting Future Floods." Also see Klinenberg, "Adaptation," 33.

7. Justin Gillis, "Are Humans to Blame? Science Is Out," *New York Times*, November 1, 2012.

8. Alison Bowen, "Local Pols Lambast Delayed Sandy Aid," *Metro New York*, January 3, 2013.

9. Bill Weinberg, "Notes from the Disaster Zone: A Survivor Sounds Off." *The Villager*, November 8–15, 2012.

10. Klinenberg, "Adaptation," 33.

11. Sheri Fink, "Where Fear, Death and Myth Collided: After Storm, Web Rumors Overtook Agony at Queens High Rise," *New York Times*, December 20, 2012.

12. Anemona Hartocollis and Julie Turkewitz, "Storm Victims in Cleanup, Face Rise in Injuries and Illness," *New York Times*, November 20, 2012.

13. David M. Halbfinger, Charles V. Bagli, and Sarah Maslin Nir, "On Ravaged Coastline, It's Rebuild Deliberately vs. Rebuild Now," *New York Times*, December 22, 2012.

14. Vivian Yee, "Fear and Anxiety amid Move to Raze Damaged Homes," *New York Times*, November 19, 2012.

15. Winnie Hu, "Fuel Rationing in City Will Go On until Friday," *New York Times*, November 19, 2012.

16. Klinenberg, "Adaptation," 32.

17. "The Power Mess on Long Island," *New York Times*, November 19, 2012.

18. Klinenberg, "Adaptation," 33.

19. Anderson, "Quinn Floats Raft of Ideas for Fighting Future Floods."

20. "Transcript of Governor Andrew M. Cuomo's 2013 State of the State Address."

Conclusion: Toward a Praxis of Cosmopolitan Citizenship

1. David Harvey, *Cosmopolitanism and the Geographies of Freedom* (New York: Columbia University Press, 2009).

selected bibliography

Abu-Lughod, Janet. *From Urban Village to East Village: The Battle for New York's Lower East Side*. Cambridge, MA: Blackwell, 1994.

Agamben, Giorgio. *Homo Sacer: Sovereign Power and Bare Life*. Trans. Daniel Heller-Roazen. Stanford, CA: Stanford University Press, 1995.

Alexander, Frederick M. *Alexander Technique: Original Writings of F.M. Alexander*. Ed. Danny McGowan. Burdett, NY: Larson, 1997.

Anderson, Benedict. *Imagined Communities: Reflections on the Origin and Spread of Nationalism*. New York: Verso, 1991.

Augé, Marc. *In the Metro*. Trans. Tom Conley. Minneapolis: University of Minnesota Press, 2002.

Bakhtin, Mikhail. *Rabelais and His World*. Trans. Helene Iswolsky. Bloomington: Indiana University Press, 1984.

Baldwin, James. *The Price of the Ticket: Collected Nonfiction, 1948–85*. New York: St. Martin's, 1985.

Barborka, Geoffrey A. *H. P. Blavatsky, Tibet and Tulku*. Adyar, India: Theosophical House, 1974.

Barnett, Jonathan. *The Fractured Metropolis: Improving the New City, Restoring the Old City, Reshaping the Region*. New York: HarperCollins, 1995.

Baudelaire, Charles. *The Painter of Modern Life and Other Essays*. London: Phaidon, 1995.

Beard, Rick, and Leslie Cohen Berlowitz. *Greenwich Village: Culture and Counterculture*. New Brunswick, NJ: Rutgers University Press, 1993.

Bender, Thomas. *The Unfinished City: New York and the Metropolitan Idea*. New York: New York University Press, 2007.

Benjamin, Gerald, and Richard P. Nathan. *Regionalism and Realism: A Study of Governments in the New York Metropolitan Area*. Washington, DC: Brookings Institution, 2001.

Benjamin, Walter. *Reflections: Essays, Aphorisms, Autobiographical Writings*. New York: Schocken, 1978.

Berger, Joseph. *The World in a City: Traveling the Globe through the Neighborhoods of the New New York*. New York: Ballantine, 2007.

Berlin, Ira, and Leslie Harris, eds. *Slavery in New York*. New York: New Press, 2005.

Berman, Marshall, and Brian Berger, eds. *New York Calling: From Blackout to Bloomberg*. London: Reaktion, 2007.

Boal, Augusto. *Theater of the Oppressed*. New York: Urizen, 1979.

Bogart, Michele H. *Public Sculpture and the Civic Ideal in New York City, 1890–1930*. Chicago: University of Chicago Press, 1989.

Braudel, Fernand. *The Mediterranean and the Mediterranean World in the Age of Philip II*. Berkeley: University of California Press, 1995.

Brecht, Bertolt. *Brecht on Theatre: The Development of an Aesthetic*. London: Methuen Drama, 1964.

Brown, Garrett Wallace, and David Held. *The Cosmopolitanism Reader*. Cambridge, MA: Polity, 2010.

Burns, Ric, James Sanders, and Lisa Ades. *New York: An Illustrated History*. New York: Knopf Doubleday, 2003.

Burrows, Edwin G., and Mike Wallace. *Gotham: A History of New York City to 1898*. New York: Oxford University Press, 2009.

Butler, Judith. *Precarious Life: The Power of Mourning and Violence*. New York: Verso, 2004.

Cantwell, Anne-Marie, and Diana diZerega Wall. *Unearthing Gotham: The Archaeology of New York City*. New Haven, CT: Yale University Press, 2001.

Caro, Robert A. *The Power Broker: Robert Moses and the Fall of New York*. New York: Random House, 1975.

Cham, Mbye B., and Claire Andrade-Watkins, eds. *Blackframes: Critical Perspectives on Black Independent Cinema*. Cambridge, MA: MIT Press, 1988.

Clifford, James, and Vivek Dhareshwar, eds. *Traveling Theories, Traveling Theorists*. Santa Cruz, CA: Inscriptions, the journal of The Center for Cultural Studies, 1989.

Cohen, Paul E., and Robert T. Augustyn. *Manhattan in Maps: 1527–1995*. New York: Rizzoli, 2006.

Cronin, John, and Robert Kennedy. *The Riverkeepers: Two Activists Fight to Reclaim Our Environment as a Basic Human Right*. New York: Touchstone, 1999.

Cudahy, Brian. *Around Manhattan Island and Other Tales of Maritime New York*. New York: Fordham University Press, 1997.

Davies, David William. *A Primer of Dutch Seventeenth Century Overseas Trade*. The Hague: Martinus Nijhoff, 1961.

Day, Leslie. *Field Guide to the Natural World of New York City*. Baltimore: Johns Hopkins University Press, 2007.

De Certeau, Michel. *The Practice of Everyday Life*. Berkeley: University of California Press, 1984.

De Laet, Johannes. *Extracts from the New World or a Description of the West Indies*. Ithaca, NY: Cornell University Library Digital Collections, 1993.

Deleuze, Gilles, and Félix Guattari. *A Thousand Plateaus: Capitalism and Schizophrenia*. Trans. Brian Massumi. Minneapolis: University of Minnesota, 1987.

Derrida, Jacques. *On Cosmopolitanism and Forgiveness*. New York: Routledge, 2001.

De Rasieres, Isaack. *New Netherland in 1627*. Ithaca, NY: The Cornell University Library Digital Collections, 1993.

De Vries, David Pietersz. "Korte Historiael Ende Journaels Aenteyckeninge," 1633–1643 (1655), in J. Franklin Jameson, Editor. *Narratives of New Netherland: 1609–1664*. New York: Elibron Classics, 2005.

———. "Voyages from Holland to America, A.D., 1633 to 1644," in New-York Historical Society, *Collections of the New-York Historical Society. Volume 111–Part 1*. New York: D. Appleton and Company, 1857.

Echenoz, Jean. *Running: A Novel*. New York: New Press, 2009.

Fanon, Frantz. *Black Skin, White Masks*. New York: Grove, 1967.

Feldenkrais, Moshé. *Awareness through Movement: Health Exercises for Personal Growth*. New York: Harper and Row, 1977.

Flint, Anthony. *Wrestling with Moses: How Jane Jacobs Took On New York's Master Builder and Transformed the American City*. New York: Random House, 2009.

Flusser, Vilém. *Towards a Philosophy of Photography*. London: Reaktion, 2000.

Foucault, Michel. *"Society Must Be Defended": Lectures at the Collège de France, 1975–1976*. Trans. David Macey. New York: Picador, 2003.

Frampton, Kenneth, and John Cava, eds. *Studies in Tectonic Culture: The Poetics of Construction in Nineteenth and Twentieth Century Architecture*. Cambridge, MA: MIT Press, 1995.

Gans, Herbert J. *The Levittowners: Ways of Life and Politics in a New Suburban Community*. New York: Random House, 1967.

Garreau, Joel. *Edge City: Life on the New Frontier*. New York: Doubleday, 1991.

Geremek, Bronislaw. *Poverty: A History*. Trans. Agnreszka Kolakowska. Cambridge, MA: Blackwell, 1994.

Gilroy, Paul. *There Ain't No Black in the Union Jack: The Cultural Politics of Race and Nation*. Chicago: University of Chicago Press, 1991.

Girard, René. *The Scapegoat*. Trans. Yvonne Freccero. Baltimore: Johns Hopkins University Press, 1986.

Glen, Susan L. *Images of America: Governors Island, NY*. Charleston, SC: Arcadia, 2009.

Goffman, Erving. *The Presentation of Self in Everyday Life*. New York: Anchor, 1959.

Gosselink, Martine. *New York New Amsterdam: The Dutch Origins of Manhattan*. New Amsterdam: Nieuw Amsterdam, 2009.

Gottman, Jean. *Megalopolis: The Urbanized Northeastern Seaboard of the United States*. Cambridge, MA: MIT Press, 1964.

Greenberg, Miriam. *Branding New York: How a City in Crisis Was Sold to the World*. New York: Routledge, 2008.

Grosz, Elizabeth. *Volatile Bodies: Toward a Corporeal Feminism*. Bloomington: Indiana University Press, 1994.

Gurdjieff, G. I. *Meetings with Remarkable Men*. New York: Dutton, 1963.

Hammett, Jerilou, and Kingsley Hammett, eds. *The Suburbanization of New York: Is the World's Greatest City Becoming Just Another Town?* New York: Princeton Architectural Press, 2007.

Hannerz, Ulf. *Exploring the City: Inquiries toward an Urban Anthropology*. New York: Columbia University Press, 1983.

———. *Transnational Connections: Culture, People, Places*. New York: Routledge, 1996.

Hansen, Joyce, and Gary McGowan. *Breaking Ground, Breaking Silence: The Story of New York's African Burial Ground*. New York: Holt, 1998.

Hartman, Saidiya V. *Scenes of Subjection: Terror, Slavery, and Self-Making in Nineteenth-Century America*. New York: Oxford University Press, 1997.

Heidegger, Martin. *Poetry, Language, Thought*. Trans. Albert Hofstadter. New York: Harper and Row, 1971.

Held, David. *Cosmopolitanism: Ideas and Realities*. Cambridge: Polity, 2010.

Hiss, Tony. *The Experience of Place: A New Way of Looking at and Dealing with Our Radically Changing Cities and Countryside*. New York: Vintage, 1991.

Ho Chi Minh. *Ho Chi Minh on Revolution: Selected Writings, 1920–1966*. Ed. Bernard Fall. New York: Praeger, 1967.

Jackson, Kenneth T., ed. *The Encyclopedia of New York City*. New Haven, CT: Yale University Press, 1995.

Jacobs, Jaap. *The Colony of New Netherland: A Dutch Settlement in Seventeenth-Century America*. Ithaca, NY: Cornell University Press, 2009.

Jacobs, Jane. *The Death and Life of Great American Cities*. New York: Random House, 1961.

———. *The Nature of Economies*. New York: Vintage, 2000.

Jameson, J. Franklin. Editor. *Narratives of New Netherland, 1609–1664*. New York: Elibron Classics, 2005.

Johnson, James Weldon. *Black Manhattan*. New York: De Capo, 1991.

Joyce, James. *Ulysses*. New York: Penguin, 2000.

Kant, Immanuel. *Toward Perpetual Peace and Other Writings on Politics, Peace, and History*. New Haven, CT: Yale University Press, 2006.

Kasinitz, Philip. *Metropolis: Center and Symbol of Our Times*. New York: New York University Press, 1995.

Kelling, George L., and Catherine M. Coles. *Fixing Broken Windows: Restoring Order and Reducing Crime in Our Communities*. New York: Free Press, 1998.

King, Anthony D., ed. *Culture, Globalization, and the World-System: Contemporary Conditions for the Representation of Identity*. Minneapolis: University of Minnesota Press, 1997.

Koeppel, Gerard T. *Water for Gotham: A History*. Princeton, NJ: Princeton University Press, 2000.

Koolhaas, Rem. *Delirious New York: A Retroactive Manifesto for Manhattan*. New York: Monacelli, 1997.

Kornblum, William. *At Sea in the City: New York from the Water's Edge*. Chapel Hill, NC: Algonquin, 2002.

Kurlansky, Mark. *The Big Oyster: History on the Half Shell*. New York: Random House, 2007.

———. *Salt: A World History*. New York: Penguin, 2003.

Lacouture, Jean. *Ho Chi Minh: A Political Biography*. New York: Random House, 1968.

Lake, Tom. *Hudson River E-Almanac*. New York State Department of Environmental Conservation, 2007–2009.

Lefebvre, Henri. *Everyday Life in the Modern World*. Trans. Sacha Rabinovitch. New Brunswick, NJ: Transaction, 1984.

Levinas, Emmanuel. *Beyond the Verse: Talmudic Readings and Lectures*. New York: Continuum, 2007.

Lewis, Tom. *The Hudson: A History*. New Haven, CT: Yale University Press, 2007.

Lopate, Phillip. *Waterfront: A Walk around Manhattan*. New York: Anchor, 2005.

Lynch, Kevin. *The Image of the City*. Cambridge, MA: MIT Press, 1960.

Macatamney, Hugh. *Cradle Days of New York*. New York: Drew and Lewis, 1909.

Marshall, Victoria, Brian McGrath, Joel Towers, and Richard Plunz, eds. *Designing Patch Dynamics*. New York: Columbia University GSAPP, 2008.

McGrath, Brian, and Jean Gardner. *Cinemetrics: Architectural Drawing Today*. Hoboken, NJ: Wiley-Academy, 2007.

McKay, Claude. *Home to Harlem*. Boston: Northeastern University Press, 1987.

Megapolensis, Johannes. *A short sketch of the Mohawk Indians in New Netherland: their land, stature, dress, manners, and magistrates, written in the year 1644, by Johannes Megapolensis, and notes, by John Romeyn Brodhead*. Ithaca, NY: Cornell University Library Digital Collections, 1993.

Montanus, Arnoldus. *The New and Unknown World: Descriptions of America and the South Lands*. Koninklijke Bibliotek, the Hague, Netherlands, 185 B 14.

Morris, Jan. *The Great Port: A Passage through New York*. New York: Oxford University Press, 1985.

Murakami, Haruki. *What I Talk about When I Talk about Running: A Memoir*. New York: Knopf, 2008.

New York City Department of City Planning. *Zoning Handbook*. New York, 2009.

New-York Historical Society. *Collections of the New-York Historical Society. Volume 111–Part 1*. New York: D. Appleton and Company, 1857.

Nieuhoff, J. *Voyages and Travels into Brazil and the East-Indies, 1640–1649*. New Delhi: Asian Educational Services, 2003.

O'Callaghan, Edmund Bailey. *History of New Netherland or New York under the Dutch. Vol. 1*. New York: Elibron Classics, 2005.

O'Neill, Joseph. *Netherland*. New York: Pantheon, 2008.
Ong, Aihwa. *Buddha Is Hiding: Refugees, Citizenship, the New America*. Berkeley: University of California Press, 2003.
Owen, David. *Green Metropolis: Why Living Smaller, Living Closer, and Driving Less Are the Keys to Sustainability*. New York: Riverhead, 2009.
Panetta, Roger. *Dutch New York: The Roots of Hudson Valley Culture*. Yonkers, NY: Hudson River Museum / Fordham University Press, 2009.
Pirenne, Henri. *Medieval Cities: Their Origins and the Revival of Trade*. Princeton, NJ: Princeton University Press, 1952.
Puar, Jasbir. *Terrorist Assemblages: Homonationalism in Queer Times*. Durham, NC: Duke University Press, 2007.
Rabelais, Francois. *Gargantua and Pantagruel*. New York: Penguin Books, 2006.
Rancière, Jacques. *The Future of the Image*. Trans. Gregory Elliot. New York: Verso, 2007.
Richman, Irwin. *Images of America: Castle Garden and Battery Park, NY*. Charleston, SC: Arcadia, 2007.
Rink, Oliver A. *Holland on the Hudson: An Economic and Social History of Dutch New York*. Ithaca, NY: Cornell University Press, 1986.
Robbins, Liz. *A Race Like No Other: 26.2 Miles through the Streets of New York*. New York: HarperCollins, 2008.
Robin, Corey. *Fear: The History of a Political Idea*. New York: Oxford University Press, 2006.
Sanderson, Eric W. *Mannahatta: A Natural History of New York City*. New York: Abrams, 2009.
Sassen, Saskia. *The Global City: New York, London, Tokyo*. Princeton, NJ: Princeton University Press, 2001.
Schechner, Richard. *Environmental Theater*. New York: Applause, 1994.
Scott, Allen J., and Edward W. Soja, eds. *The City: Los Angeles and Urban Theory at the End of the Twentieth Century*. Berkeley: University of California Press, 1996.
Seitz, Sharon, and Stuart Miller. *The Other Islands of New York City: A History and Guide*. Woodstock, VT: Countryman, 1996.
Sennett, Richard. *The Fall of Public Man*. New York: W. W. Norton, 1992.
———. *Flesh and Stone: The Body and the City in Western Civilization*. New York: W. W. Norton, 1996.
Shank, Theodore. *American Alternative Theater*. New York: Grove, 1982.
Shorto, Russell. *The Island at the Center of the World: The Epic Story of Dutch Manhattan and the Forgotten Colony That Shaped America*. New York: Doubleday, 2004.
Smith, William Edgett. *Nyerere of Tanzania*. Harare: Zimbabwe Publishing House, 1981.
Soja, Edward W. *Postmetropolis: Critical Studies of Cities and Regions*. Malden, MA: Blackwell, 2000.

———. *Thirdspace: Journeys to Los Angeles and Other Real-and-Imagined Places*. Cambridge, MA: Blackwell, 1996.

Sorkin, Michael, and Sharon Zukin, eds. *After the World Trade Center: Rethinking New York City*. New York: Routledge, 2002.

Srinivas, Smriti. *Landscapes of Urban Memory: The Sacred and the Civic in India's High-Tech City*. Minneapolis: University of Minnesota Press, 2001.

Staden, Hans. *Hans Staden's True History: An Account of Cannibal Captivity in Brazil*. Ed. and trans. Neil L. Whitehead and Michael Harbsmeier. Durham, NC: Duke University Press, 2008.

Stallybrass, Peter, and Allon White. *The Politics and Poetics of Transgression*. Ithaca, NY: Cornell University Press, 1986.

Taussig, Michael. *Shamanism, Colonialism, and the Wild Man: A Study in Terror and Healing*. Chicago: University of Chicago Press, 1991.

The Theosophical Movement, 1875–1950. Los Angeles: Cunningham, 1951.

Thompson, Robert Farris. *Flash of the Spirit: African and Afro-American Art and Philosophy*. New York: Random House, 1984.

Tönnies, Ferdinand. *Community and Society*. New Brunswick, NJ: Transaction, 1988.

Tschumi, Bernard. *Architecture and Disjunction*. Cambridge, MA: MIT Press, 1996.

———. *Event-Cities*. Cambridge, MA: MIT Press, 1994.

Turner, Victor. *Dramas, Fields, and Metaphors: Symbolic Action in Human Society*. Ithaca, NY: Cornell University Press, 1975.

Van Der Donck, Adriaen. *A Description of New Netherland*. Lincoln: University of Nebraska, 2008.

Van Linschoten, Jan Huygen. *The Voyage to the East Indies*. Volumes 1 and 2. London: Elibron Classics, 2005.

Vingboons, Johannes. *Atlas van kaarten en aanzichten van de VOC en WIC, genoemd Vingboons-atlas, in Het algemeen rijksarchief te's-Gravenhage*. [Bussum]: Fibula-Van Dishoeck, 1981.

Virga, Vincent. *Historic Maps and Views of New York*. New York: Black Dog and Leventhal, 2008.

Virilio, Paul. *Speed and Politics*. Trans. Marc Polizzotti. Los Angeles: Semiotext(e), 2007.

Virno, Paolo. *A Grammar of the Multitude: For an Analysis of Contemporary Forms of Life*. Trans. Isabella Bertoletti and James Cascaito. Los Angeles: Semiotext(e), 2009.

Vitale, Alex. *City of Disorder: How the Quality of Life Campaign Transformed New York Politics*. New York: New York University Press, 2009.

Waldman, Stuart, and Zack Winestine. *Maritime Mile: The Story of the Greenwich Village Waterfront*. New York: Mikaya, 2002.

Wall, Diana diZerega, and Anne-Marie Cantwell. *Touring Gotham's Archaeological Past: 8 Self-Guided Walking Tours through New York City*. New Haven, CT: Yale University Press, 2004.

Weber, Max. *The City*. New York: Free Press, 1968.

Wetzsteon, Ross. *Republic of Dreams: Greenwich Village: The American Bohemia, 1910–1960*. New York: Simon and Schuster, 2002.

Wiley and Putnam's Emigrant's Guide: Comprising Advice and Instruction in Every Stage of the Voyage to America. London: Wiley and Putnam, 1845.

index